Rigidity Theorems for Actions of Product Groups and Countable Borel Equivalence Relations

MEMOIRS
of the
American Mathematical Society

Number 833

Rigidity Theorems for Actions of Product Groups and Countable Borel Equivalence Relations

Greg Hjorth
Alexander S. Kechris

September 2005 • Volume 177 • Number 833 (first of 4 numbers) • ISSN 0065-9266

American Mathematical Society
Providence, Rhode Island

2000 *Mathematics Subject Classification.*
Primary 03E15, 28D15, 37A15, 37A20.

Library of Congress Cataloging-in-Publication Data

Hjorth, Greg, 1963–
Rigidity theorems for actions of product groups and countable Borel equivalence relations / Greg Hjorth, Alexander S. Kechris.
p. cm. — (Memoirs of the American Mathematical Society, ISSN 0065-9266 ; no. 833)
"Volume 177, number 833 (first of 4 numbers)."
Includes bibliographical references.
ISBN 0-8218-3771-0 (alk. paper)
1. Set theory. 2. Borel sets. 3. Measure theory. 4. Ergodic theory. 5. Rigidity (Geometry) I. Kechris, A. S., 1946– II. Title. III. Series.

QA3.A57 no. 833
[QA248]
510 s—dc22
[511.3′22] 2005044096

Memoirs of the American Mathematical Society

This journal is devoted entirely to research in pure and applied mathematics.

Subscription information. The 2005 subscription begins with volume 173 and consists of six mailings, each containing one or more numbers. Subscription prices for 2005 are $606 list, $485 institutional member. A late charge of 10% of the subscription price will be imposed on orders received from nonmembers after January 1 of the subscription year. Subscribers outside the United States and India must pay a postage surcharge of $31; subscribers in India must pay a postage surcharge of $43. Expedited delivery to destinations in North America $35; elsewhere $130. Each number may be ordered separately; *please specify number* when ordering an individual number. For prices and titles of recently released numbers, see the New Publications sections of the *Notices of the American Mathematical Society.*

Back number information. For back issues see the *AMS Catalog of Publications.*

Subscriptions and orders should be addressed to the American Mathematical Society, P. O. Box 845904, Boston, MA 02284-5904, USA. *All orders must be accompanied by payment.* Other correspondence should be addressed to 201 Charles Street, Providence, RI 02904-2294, USA.

Memoirs of the American Mathematical Society is published bimonthly (each volume consisting usually of more than one number) by the American Mathematical Society at 201 Charles Street, Providence, RI 02904-2294, USA. Periodicals postage paid at Providence, RI. Postmaster: Send address changes to Memoirs, American Mathematical Society, 201 Charles Street, Providence, RI 02904-2294, USA.

This publication is indexed in *Science Citation Index®*, *SciSearch®*, *Research Alert®*, *CompuMath Citation Index®*, *Current Contents®/Physical, Chemical & Earth Sciences.*
Printed in the United States of America.

∞ The paper used in this book is acid-free and falls within the guidelines established to ensure permanence and durability.
Visit the AMS home page at `http://www.ams.org/`

10 9 8 7 6 5 4 3 2 1 10 09 08 07 06 05

Contents

Abstract

This Memoir is both a contribution to the theory of Borel equivalence relations, considered up to Borel reducibility, and measure preserving group actions considered up to orbit equivalence. Here E is said to be *Borel* reducible to F if there is a Borel function f with xEy if and only if $f(x)Ff(y)$. Moreover, E is *orbit equivalent* to F if the respective measure spaces equipped with the extra structure provided by the equivalence relations are almost everywhere isomorphic.

We consider product groups acting ergodically and by measure preserving transformations on standard Borel probability spaces. In general terms, the basic parts of the monograph show that if the groups involved have a suitable notion of "boundary" (we make this precise with the definition of *near hyperbolic*), then one orbit equivalence relation can only be Borel reduced to another if there is some kind of algebraic resemblance between the product groups and coupling of the action. This also has consequence for orbit equivalence. In the case that the original equivalence relations do not have non-trivial almost invariant sets, the techniques lead to *relative ergodicity* results. An equivalence relation E is said to be *relatively ergodic* to F if any f with $xEy \Rightarrow f(x)Ff(y)$ has $[f(x)]_F$ constant almost everywhere.

This underlying collection of lemmas and structural theorems is employed in a number of different ways.

One of the most pressing concerns was to give completely self-contained proofs of results which had previously only been obtained using Zimmer's superrigidity theory. We present "elementary proofs" that there are incomparable countable Borel equivalence relations (Adams-Kechris), inclusion does not imply reducibility (Adams), and $(n+1)E$ is not necessarily reducible to nE (Thomas).

In the later parts of the paper we give applications of the theory to specific cases of product groups. In particular, we catalog the actions of products of the free group and obtain additional rigidity theorems and relative ergodicity results in this context.

There is a rather long series of appendices, whose primary goal is to give the reader a comprehensive account of the basic techniques. But included here are also some new results. For instance, we show that the Furstenberg-Zimmer lemma on cocycles from amenable groups fails with respect to Baire category, and use this to answer a question of Weiss. We also present a different proof that F_2 has the Haagerup approximation property.

Received by the editor December 2, 2002.

2000 *Mathematics Subject Classification.* Primary 03E15, 28D15, 37A15, 37A20.

Key words and phrases. Borel equivalence relations, ergodic theory of nonamenable groups, product group actions, rigidity, Borel reducibility.

The first author was supported in part by NSF Grant DMS-9970403, DMS-0140503.

The second author was supported in part by NSF Grant DMS-9987437.

Introduction

A) This paper is a contribution to the theory of countable Borel equivalence relations on standard Borel spaces. As usual, by a *standard Borel space* we mean a Polish (complete separable metric) space equipped with its σ-algebra of Borel sets. An equivalence relation E on a standard Borel space X is *Borel* if it is a Borel subset of X^2.

Given two equivalence relations E, F on spaces X, Y, resp., we say that E is *Borel reducible* to F, in symbols,

$$E \leq_B F,$$

if there is a Borel map $\rho : X \to Y$ such that

$$xEy \Leftrightarrow \rho(x) F \rho(y).$$

We also say that E is *Borel bireducible* to F, in symbols

$$E \sim_B F$$

if $E \leq_B F$ and $F \leq_B E$. Finally put

$$E <_B F \Leftrightarrow E \leq_B F \ \& \ F \not\leq_B E.$$

We refer the reader to [Ke99] for a detailed discussion of the motivation for the study of the reducibility order on Borel (and more general definable) equivalence relations. On the one hand, this can be viewed as providing the basic underlying concept for the development of a theory of complexity of classification problems in mathematics. On the other hand, it can be understood as the basis of a theory of Borel (as opposed to the classical Cantor) cardinality of quotients of standard Borel spaces by Borel equivalence relations. One can view here $E \leq_B F$ as expressing that X/E has Borel cardinality $\leq$ to that of Y/F, and $E \sim_B F$ as expressing that $X/E, Y/F$ have the same Borel cardinality.

In this paper, we will only discuss countable Borel equivalence relations. An equivalence relation E on a space X is called *countable* if every equivalence class $[x]_E, x \in X$, is countable. These include many important examples, like, for instance, all equivalence relations induced by Borel actions of countable (discrete) groups on standard Borel spaces, and, up to bireducibility, even those induced by Borel actions of Polish locally compact groups, the isomorphism relation on various types of countable structures that have "finite type" (again this is up to bireducibility), Turing or arithmetical equivalence, etc. Their study is actually closely connected with ergodic theory and other areas in dynamical systems, since by a result of Feldman-Moore [FM], the countable Borel equivalence relations are in fact exactly those induced by Borel actions of countable groups on standard Borel spaces.

Let us note here that, when E, F are countable Borel equivalence relations, $E \sim_B F$ can be also expressed as follows: $E \sim_B F$ iff there are Borel sets $A \subseteq X, B \subseteq Y$ meeting every E, F-class, resp., and a Borel isomorphism of $E|A$ with $F|B$. Also $E \sim_B F$ is equivalent to the existence of a bijection $\sigma : X/E \rightarrow Y/F$ which is "Borel", in the sense that it admits a Borel lifting (in both directions), i.e., there are Borel maps $\rho : X \rightarrow Y, \rho' : Y \rightarrow X$ such that $[\rho(x)]_F = \sigma([x]_E), [\rho'(y)]_E = \sigma^{-1}([y]_F)$, for $x \in X, y \in Y$ (see [DJK], 2.6).

B) We first state some basic facts concerning the structure of $\leq_B$ on countable Borel equivalence relations, for which we refer the reader to [DJK], [JKL]. Below we identify a standard Borel space X with the equivalence relation $\Delta(X)$ of equality on X, and we call a countable Borel equivalence relation *tame* if it admits a Borel selector. Finally, we call a Borel equivalence relation *finite* if every equivalence class is finite (so a finite relation is tame), and *hyperfinite* if it can be written as a union of an *increasing* sequence of finite Borel equivalence relations. We now have:

(i) $1 <_B 2 <_B 3 <_B \cdots <_B n <_B \cdots <_B \mathbb{N} <_B \mathbb{R}$ (where n is any space of cardinality n) and, up to $\sim_B$, these are exactly the tame countable Borel equivalence relations.

(ii) All non-tame hyperfinite Borel equivalence relations are Borel bireducible to each other, and if E_0 denotes any one of them, then

$$1 <_B 2 <_B \cdots <_B \mathbb{N} <_B \mathbb{R} <_B E_0$$

is an initial segment of the countable Borel equivalence relations under $\leq_B$.

(iii) There is a largest countable Borel equivalence relation, in the sense of $\leq_B$, called *universal* and denoted by E_∞. It is of course unique, up to $\sim_B$. Moreover, $E_0 <_B E_\infty$. Thus every countable Borel equivalence relation is, up to $\sim_B$, one of $1, 2, \ldots, n, \ldots, \mathbb{N}, \mathbb{R}$, if it is smooth, or else belongs in the interval $E_0 \leq E \leq E_\infty$.

It has been also known for quite some time that there are *intermediate* equivalence relations

$$E_0 <_B E <_B E_\infty,$$

but until very recently only a small finite number of distinct, up to $\sim_B$, examples were known and they were all comparable under $\leq_B$. Then Adams-Kechris [AK] showed that the structure of $\leq_B$ on countable Borel equivalence relations is indeed complex, by establishing that the partial order of inclusion among Borel sets of reals can be embedded in $\leq_B$ on countable Borel equivalence relations, so, for example, any Borel partial order can be embedded as well. They also showed that the relations $\leq_B, \sim_B$ on countable Borel equivalence relations are $\mathbf{\Sigma}^1_2$-complete (in the codes).

C) The proofs of the main results of [AK] made heavy use of work of Zimmer [Zi84] in the ergodic theory of linear algebraic groups, in particular the so-called *superrigidity theory.* The key point can be informally summarized by saying that there is in our context a phenomenon of *set theoretic rigidity* analogous to the *measure theoretic rigidity* discovered by Zimmer.

Roughly speaking, measure theoretic rigidity refers to the fact that, under certain circumstances, the equivalence relation induced by a group action, with an associated invariant measure, "encodes" or "remembers" a lot of information about the group (and the action). Set theoretic rigidity, refers to the fact that such information about the group is simply "encoded" in the Borel cardinality of the orbit space (i.e., the quotient space of the induced equivalence relation). For example, if we use the notation $|\ |_B$ to denote Borel cardinalities, then for the standard action of $\mathrm{GL}_n(\mathbb{Z})$ (= the group of $n \times n$ integer matrices with determinant ± 1) by matrix multiplication on $\mathbb{T}^n$, we have: $m < n \Rightarrow |\mathbb{T}^m/\mathrm{GL}_m(\mathbb{Z})|_B < |\mathbb{T}^n/\mathrm{GL}_n(\mathbb{Z})|_B$, so that, in particular, we have invariance of dimension, i.e., the Borel cardinality of $\mathbb{T}^n/\mathrm{GL}_n(\mathbb{Z})$ "encodes" the dimension n. (Of course the classical cardinality of these quotient spaces is that of the continuum, for each n.) Also, if for a countable group Γ we consider the shift action of Γ on 2^Γ ($\gamma \cdot f(\delta) = f(\gamma^{-1}\delta)$, for $\gamma \in \Gamma, f \in 2^\Gamma$), and denote by $(2)^\Gamma = \{f \in 2^\Gamma : \forall \gamma \neq 1(\gamma \cdot f \neq f)\}$ the *free part* of the action, and by $F(\Gamma, 2)$ the induced equivalence relation on $(2)^\Gamma$, then for $\Gamma_p = \mathrm{SO}_7(\mathbb{Z}[1/p])$, p a prime, we have $F(\Gamma_p, 2) \leq_B F(\Gamma_q, 2) \Leftrightarrow p = q$, so $F(\Gamma_p, 2)$ "encodes" p, and in particular the equivalence relations $F(\Gamma_p, 2)$ are incomparable under $\leq_B$. For some very interesting more recent applications of these ideas to the classical classification problem of finite rank torsion-free abelian groups, see the papers of S. Thomas [T01] (and the references contained therein) and [T02].

D) It is therefore important to understand further the phenomenon of set theoretic rigidity as well as its connection with measure theoretic rigidity. To start with, there is an important dividing line, concerning rigidity, that of amenability versus non-amenability. Amenability is associated with "elasticity" (the opposite of rigidity) and non-amenability is (often) associated to rigidity. Recall that a countable group is *amenable* if it admits an invariant finitely additive probability measure defined on all its subsets. For example, all abelian, solvable, etc., groups are amenable, but all countable groups containing free groups $F_n, 2 \leq n \leq \infty$, are not.

In the sequel, we will use the following notation: If Γ acts on X, the induced equivalence relation will be denoted by E_Γ^X, so that

$$xE_\Gamma^X y \Leftrightarrow \exists \gamma \in \Gamma(\gamma \cdot x = y).$$

We will also use the convention that *measure* on a standard Borel space always means *Borel probability measure*. Finally, we recall the following terminology from ergodic theory: If E, F are countable Borel equivalence relations on standard Borel spaces X, Y, resp., and μ, ν are measures on X, Y, resp., then we say that E is *orbit equivalent* (OE) to F iff there are Borel co-null sets A, B in X, Y resp., which are E, F-invariant, resp., and a Borel bijection $\pi : A \to B$ which sends E to F and μ to ν (i.e., for $x, y \in A, xEy \Leftrightarrow \pi(x)F\pi(y)$, and $\pi_*\mu = \nu$, where $\pi_*\mu(Y_0) = \mu(\pi^{-1}(Y_0))$, for any Borel $Y_0 \subseteq Y$). We say that E is *stably orbit equivalent* to F (SOE) if there are Borel sets $A \subseteq X, B \subseteq Y$ of positive measure, with A meeting almost every E-class and B meeting almost every F-class, such that if μ_A, ν_B, resp., denote the normalized restrictions of μ, ν to A, B, resp., then there is a Borel bijection $\pi : A \to B$ sending E to F and μ_A to ν_B. Thus SOE is the measure theoretic analog of bi-reducibility, $\sim_B$. We also say that Borel actions of countable groups Γ, Δ on X, Y, which carry measures μ, ν, resp., are OE or SOE if the corresponding E_Γ^X, E_Δ^Y are OE or SOE.

Now Ornstein-Weiss [OW] have shown that if Γ is amenable and acts in a Borel way on a standard Borel space X, then for every measure μ, E_Γ^X is μ-hyperfinite, i.e., hyperfinite when restricted to a conull Borel Γ-invariant set. From this and Dye's Theorem [D], one sees, for example, that if two countable amenable groups Γ_1, Γ_2 act freely on standard Borel spaces X_1, X_2 (an action being *free* if $\gamma \cdot x \neq x, \forall \gamma \neq 1$) with invariant, ergodic measures μ_1, μ_2, then $E_{\Gamma_1}^{X_1}$ is OE to $E_{\Gamma_2}^{X_2}$ (ergodicity of course means that Γ_i-invariant Borel sets are either null or conull). Conversely, if Γ_1, Γ_2 are arbitrary countable groups, and Γ_1 acts freely on X_1 with invariant measure μ_1, while Γ_2 acts freely on X_2 with invariant measure μ_2, then if $E_{\Gamma_1}^{X_1}$ is SOE to $E_{\Gamma_2}^{X_2}$, Γ_1 is amenable iff Γ_2 is amenable. Thus for an amenable group Γ, acting freely on X with ergodic, invariant measure μ, (E_Γ^X, μ) "encodes" nothing, except the amenability of Γ, thereby exhibiting the "elasticity" that we referred to earlier.

In the set theoretic context, this "elasticity" is only partially known at the present time. It was shown in [JKL], following work of Weiss for the groups $\mathbb{Z}^n$, that if Γ is a finitely generated group of polynomial growth, then for any Borel action of Γ on a standard Borel space X, E_Γ^X is hyperfinite, and thus if Γ_1, Γ_2 are two such groups acting on X_1, X_2 so that $E_{\Gamma_1}^{X_1}, E_{\Gamma_2}^{X_2}$ are not tame, then $E_{\Gamma_1}^{X_1} \sim_B E_{\Gamma_2}^{X_2} (\sim_B E_0)$.

For example, one cannot tell apart non-tame actions of $\mathbb{Z}^m, \mathbb{Z}^n$, up to $\sim_B$, for any m, n.

It is open however whether the previous result extends to *all* amenable groups (an open problem raised by Weiss in [W84]).

E) In the non-amenable case, there are several important measure theoretic rigidity results for actions of lattices of simple Lie groups of higher rank, see e.g., Zimmer [Zi84] and, more recently, Furman [Fu]. We will concentrate below on results and problems concerning the simplest, in some sense, non-amenable groups, namely the free (non-abelian) groups and related ones.

F) Solving a long standing problem, Gaboriau [Ga00] has recently shown that the idea of the "cost" of an equivalence relation, introduced in Levitt [Lev], provides a new invariant that leads to the following rigidity result: Let the free groups F_m, F_n act freely in a Borel way on standard Borel spaces X, Y resp., with invariant probability measures μ, ν, resp. Then if $E_{F_m}^X$ is OE to $E_{F_n}^Y$, we have $m = n$. (Concerning actions of the same free group, say F_2, it is known that there are non-OE measure preserving, ergodic free Borel actions of F_2, but only finitely many distinct examples are known at this time (see the discussion in §1, **C**) below).)

However, it is well-known that this fails if OE is replaced by SOE, which is the measure theoretic analog of bireducibility, $\sim_B$. For example, there are free, ergodic measure preserving actions of F_2, F_3 which are SOE.

One can therefore ask whether there is any set theoretic rigidity for actions of free groups, and the plain answer at this time is that we do not know. In fact, our ignorance goes much further than this.

Consider all the equivalence relations E_F^X, induced by a free Borel action of a countable free group F on a standard Borel space X. Up to $\sim_B$, these are exactly the so-called *treeable* countable Borel equivalence relations. Here E, a countable Borel equivalence relation on X, is *treeable* if there is a Borel acyclic graph on X, whose connected components are the E-classes, i.e., we can assign in a uniform Borel way to each E-class C a tree T_C with vertex set C. Among the non-tame treeable equivalence relations there is a smallest one, in the sense of $\leq_B$, namely E_0, and a largest one, called *universal treeable*, and denoted by $E_{\infty T}$, which is of course unique, up to $\sim_B$. One realization of $E_{\infty T}$, is, for example, $F(F_2, 2)$, the equivalence relation induced by the shift action of F_2 on $(2)^{F_2}$, which is the free part of the shift action of F_2 on 2^{F_2}. It is well-known that $E_0 <_B E_{\infty T}$ (see, e.g.,

[JKL]) but it is not known whether there are any E strictly between $E_0, E_{\infty T}$:

$$E_0 <_B E <_B E_{\infty T}.$$

In other words, it is not known if there are treeable non-hyperfinite Borel equivalence relations, which are different, up to $\sim_B$, than $E_{\infty T}$. In particular, since every free Borel action of a free non-abelian group F on a standard Borel space X, which admits an invariant measure, has the property that E_F^X is not hyperfinite, it is unknown whether all such E_F^X are the same, up to $\sim_B$. It is natural to conjecture that this is not the case, but we are missing the techniques to demonstrate that.

In fact, even the following seems to be open: Is $E_{\infty T}$ the smallest, in the sense of $\leq_B$, countable Borel equivalence relation which is bigger (in $<_B$) than E_0?

G) Although we are completely ignorant about set theoretic rigidity in the realm of free actions of free groups, it turns out that there are interesting such rigidity phenomena, when one considers product groups, such as $F_2 \times F_2, F_2 \times \mathbb{Z}, \dots$ (and sometimes even such products involving hyperbolic groups).

Gaboriau [Ga01] has recently introduced some new invariants, called ℓ^2-Betti numbers, ergodic dimension, etc., inspired by ideas in algebraic topology, in which one studies, instead of trees, higher dimensional analogs, i.e., contractible finite dimensional simplicial complexes. These can be used to obtain measure theoretic rigidity results as well as set theoretic ones.

In this paper, we employ a different method, which originates in work of Mostow [Mo], Margulis [Ma77, Ma77a] (see also [Zi84]) and, in a context closer to ours, further used in papers of Zimmer [Zi81] and Adams [Ad88, Ad94, Ad94a, Ad95, Ad96]. It has to do with the action of groups on "boundaries", and allows us to obtain several set theoretic (as well as measure theoretic) rigidity results for actions of product groups. An interesting feature of this method, at least as it is used in our paper, is its "elementary" character, as it requires no more than standard measure theory and functional analysis.

Although there are some results that can be obtained by both Gaboriau's methods and the methods used in our paper, these techniques seem to be complementary to each other.

H) We will not try to list in this introduction the various rigidity theorems that are obtained here, and which often take the form of cocycle reduction results, but we will rather state some of their applications, concerning the structure of $\leq_B$ on countable Borel equivalence relations, that is our main motivation here.

The first application is to give an "elementary" proof of the main results in Adams-Kechris [AK], one that avoids the use of Zimmer's Superrigidity Theory. This is based on the following result, proved in Chapter 3, **(B)**.

Below, if E is a countable Borel equivalence relation on a standard Borel space X, μ is a measure on X and F is a countable Borel equivalence relation on a standard Borel space Y, we say that (E, μ), or simply E, is *F-ergodic* if for every Borel *homomorphism* $\rho : X \to Y$ of E to F (i.e., a map satisfying $x_1 E x_2 \Rightarrow \rho(x_1) F \rho(x_2)$), there is a conull set in X which is mapped by ρ into a single F-class.

Theorem 1. *For each non-empty set S of odd primes, let*

$$\Gamma_S = (*_{p \in S}(\mathbb{Z}_p * \mathbb{Z}_p)) \times \mathbb{Z}$$

and consider the shift action of Γ_S *on* 2^{Γ_S} $(\gamma \cdot f(\delta) = f(\gamma^{-1}\delta))$ *and the equivalence relation induced by this action on its free part* $(2)^{\Gamma_S} = \{f \in 2^{\Gamma_S} : \forall \gamma \in \Gamma_S(\gamma \cdot f \neq f)\}$, *which is denoted by* $F(\Gamma_S, 2)$. *Let* μ_S *be the usual product measure on* 2^{Γ_S}, *which concentrates on* $(2)^{\Gamma_2}$. *Then*

$$S \not\subseteq T \Rightarrow E_S \text{ is } E_T\text{-ergodic.}$$

For example, if $p \neq q$ are odd primes and $\Gamma_p = \Gamma_{\{p\}}, \Gamma_q = \Gamma_{\{q\}}$, then $F(\Gamma_p, 2), F(\Gamma_q, 2)$ are incomparable under $\leq_B$. Notice that in some sense the equivalence relations $F(\Gamma_p, 2), F(\Gamma_q, 2)$ are "just above" $E_{\infty T}$, since $F(\mathbb{Z}_p * \mathbb{Z}_p, 2) \sim_B F(\mathbb{Z}_q * \mathbb{Z}_q, 2) \sim_B E_{\infty T}$ (see [JKL, §3]). In fact, replacing $\mathbb{Z}$ by an infinite locally finite group in the definition of Γ_p above, produces incomparable equivalence relations, which are unions of increasing sequences of treeable equivalence relations.

Our next application, discussed in §3, **(C)**, is to provide "elementary" proofs of two other, more recent results, in the theory of Borel reducibility, that were originally proved by using Zimmer's Superrigidity Theory as well as Ratner's measure classification theorem [Ra]. First, Adams [Ad02], in response to a question of S. Thomas, constructed the first examples of countable Borel equivalence relations E, F, on an uncountable Polish space, with $E \subseteq F$ but $E \not\leq_B F$. Thomas [T02a] then used these techniques to solve two other well-known problems, by constructing the first example of a countable Borel equivalence relation E, on an uncountable Polish space, satisfying $E <_B 2E$, where $2E$ denotes the (disjoint) sum of two copies of E, and also the first examples of aperiodic (i.e., having no finite classes) countable Borel equivalence relations E, F with $E \sim_B F$ but for which $E \approx_B F$ fails. (Here $E \approx_B F$ means that E is Borel bireducible to F via *injective reductions*.) In §3, **(C)** we give new proofs of these results, that avoid superrigidity and Ratner's theorem.

For the next result, proved in Chapter 3, **(D)**, we need the following concepts. If a countable group Γ acts in a Borel way on a standard Borel space X and μ is a Γ-invariant measure on X, then for any $\Gamma_1 \subseteq \Gamma$ we say that Γ_1 acts *ergodically* if every Γ_1-invariant Borel set in X is either null or conull. Given any countable Borel equivalence relation F on a standard Borel space Y, we say that Γ_1 acts *F-ergodically* if for every Borel homomorphism $\rho : X \to Y$ of $E^X_{\Gamma_1}$ to F, there is a conull set mapped by ρ into a single F-class. Thus Γ_1 acts ergodically iff it acts F-ergodically for any tame F, and Γ_1 acts E_0-ergodically iff it acts F-ergodically for any hyperfinite F.

Theorem 2. *Let* $n \geq 1, \Gamma = \Gamma_1 \times \cdots \times \Gamma_n \times \Delta$, *where* Γ_i *are countable groups and* Δ *is a countable infinite amenable group. Assume* Γ *acts in a Borel way on a standard Borel space* X *with invariant measure* μ, *so that* Δ *acts ergodically, and each* Γ_i *acts* E_0*-ergodically. Then if* $H = H_1 \times \cdots \times H_n$ *is any product of free groups and* H *acts freely in a Borel way on a standard Borel space* Y, *we have that* E^X_Γ *is* E^Y_H*-ergodic.*

In Chapter 5, **(B)** we also prove the following non-reducibility result, (approximately) weakening the hypothesis and the conclusion. For the notion of hyperbolic groups, see, e.g., [GdlH].

Theorem 3. *Let* $n \geq 1$ *and* $\Gamma = \Gamma_1 \times \cdots \times \Gamma_n \times \Delta$, *where each* Γ_i *is a countable non-amenable group and* Δ *is a countable infinite amenable group. Let* Γ *act freely in a Borel way on a standard Borel space* X *with invariant measure* μ *and assume* *that Δ acts ergodically. Let $H = H_1 \times \cdots \times H_n$, where each H_i is a torsion-free hyperbolic group (e.g., a free group) and let H act freely in a Borel way on a standard Borel space Y. Then*

$$E_\Gamma^X \not\leq_B E_H^Y.$$

In Chapter 5, **(A)**, we also prove some rigidity results, of which the following is a sample.

Theorem 4. *Let H_0 be a non-amenable countable torsion-free hyperbolic group (e.g., a non-abelian free group) and let Δ_0 be an infinite amenable group. Let also H_1 be a torsion-free hyperbolic group and let Δ_1 be an amenable group. Suppose $H_0 \times \Delta_0$ acts freely in a Borel way on a standard Borel space X with invariant measure μ, so that Δ_0 acts ergodically, and $H_1 \times \Delta_1$ acts freely in a Borel way on a standard Borel space Y. Then if $E^X_{H_0\times\Delta_0} \leq_B E^Y_{H_1\times\Delta_1}$, H_0 is isomorphic to a subgroup of H_1.*

There is also a version of this result for stable orbit equivalence.

Theorem 5. *In the context of Theorem 4, assume also that Δ_1 is infinite, the action of $H_1 \times \Delta_1$ on Y has an invariant measure ν and Δ_1 acts ergodically. If the action of $H_0 \times \Delta_0$ on X is SOE to the action of $H_1 \times \Delta_1$ on Y, then H_0 is isomorphic to H_1.*

Recall that there are, for example, ergodic free measure preserving actions of F_2, F_3 which are SOE. In particular, by considering product actions, there are ergodic free measure preserving actions of $F_2 \times \mathbb{Z}, F_3 \times \mathbb{Z}$ which are SOE. However, the preceding result shows that if free measure preserving actions of $F_m \times \mathbb{Z}, F_n \times \mathbb{Z}$ are $\mathbb{Z}$-ergodic, and they are SOE, then $m = n$.

We next consider product actions of groups. A sample result is the following, proved in §7. (This also follows from a recent, unpublished, result of Gaboriau, who uses very different methods.)

Theorem 6. *Let $n \geq 1$ and suppose F_2^n acts freely in a Borel way on a standard Borel space X_1 with invariant measure μ_1 and $\mathbb{Z}$ acts freely in a Borel way on a standard Borel space X_2 with invariant measure μ_2. Consider the product action of $F_2^n \times \mathbb{Z}$ on $X_1 \times X_2$ ($(\gamma_1,\gamma_2)\cdot(x_1,x_2) = (\gamma_1\cdot x_1, \gamma_2\cdot x_2)$). Then for any treeable Borel equivalence relations $E_1, \ldots, E_n$, $E^{X_1\times X_2}_{F_2^n\times\mathbb{Z}} \not\leq_B E_1 \times \cdots \times E_n$ (where $(x_1,\ldots,x_n)E_1 \times \cdots \times E_n(y_1,\ldots y_n) \Leftrightarrow \forall i \leq n(x_i E_i y_i)$).*

Finally, in Chapter 8, by putting together several of these results and also using some theorems in Gaboriau [Ga01], we obtain the following application, which among other things, completely determines the relationship between the products $(E_{\infty T})^n \sim_B E(F_2,2)^n$, i.e., products of the shift equivalence relation of F_2, and (the free part of) the shift equivalence relation of the product groups F_2^n, $F(F_2^n,2)$. Below Z is any infinite locally finite countable group (e.g., $\mathbb{Z}_2 \oplus \mathbb{Z}_2 \oplus \ldots$).

Theorem 7. i)

$$E_{\infty T} <_B E_{\infty T} \times E_0 <_B (E_{\infty T})^2 <_B \cdots <_B (E_{\infty T})^n <_B (E_{\infty T})^n \times E_0 <_B (E_{\infty T})^{n+1} <_B \ldots$$

ii)

$$E_{\infty T} <_B F(F_2 \times Z, 2) <_B F(F_2^2, 2) <_B \cdots <_B F(F_2^n, 2)$$

$$<_B F(F_2^n \times Z, 2) <_B F(F_2^{n+1}, 2) <_B \ldots$$

iii) *Finally, for* $n \geq 1, (E_{\infty T})^n \leq_B F(F_2^n, 2)$ *and* $(E_{\infty T})^n \times E_0 \leq_B F(F_2^n \times Z, 2)$, *but* $F(F_2 \times Z, 2) \not\leq_B (E_{\infty T})^n, (E_{\infty T})^n \not\leq_B F(F_2^{n-1} \times Z, 2)$ *and* $(E_{\infty T})^n \times E_0 \not\leq_B F(F_2^n, 2)$. *In particular* $F(F_2^n, 2)$ *and* $(E_{\infty T})^m$ *are incomparable in* $\leq_B$, *if* $2 \leq n < m$.

It was already known that $E_{\infty T} <_B E_{\infty T} \times E_0$ (see [JKL, 3.28]) and $(E_{\infty T})^n < E_\infty$ (see [HK, 10.8]).

The paper is organized as follows. After a chapter on preliminaries, we discuss in Chapter 1 treeable equivalence relations. In Chapter 2, we prove a cocycle reduction result concerning E_0-ergodic actions of groups. The related Appendix A discusses E_0-ergodicity and other strong notions of ergodicity, and contains proofs of related facts that are needed in this paper. It also provides a survey of relevant results in the literature. Also important here is Appendix B, which provides background information about cocycles and cocycle-invariant functions. The ideas involved in this Appendix can be also used to solve a problem of Weiss, see [W00], p. 290, concerning equivariant projections on the space of sequences of Baire measurable functions, see B3.5. In Chapter 3, we provide applications of this cocycle reduction result. Relevant here is Appendix C, which discusses in general concepts and results, concerning actions of groups on trees and boundaries, that are needed in our paper, and also summarizes some important properties of hyperbolic groups that we use here. In Chapter 4, we consider what happens when E_0-ergodicity is relaxed and prove a "factoring" result that often allows us to deal with this relaxation. Applications are given in Chapter 5. In Chapters 6, 7, 8, we discuss the case of product actions and give various further applications. The work in Chapter 7 also uses ideas and results in Gaboriau [Ga01], and the related Appendix D provides some links between concepts contained in that paper and Borel reducibility. Finally, Appendix E contains a full proof of the factoring Theorem 4.4, which is only proved in a special case in Chapter 4, sufficient for the applications in Chapter 5.

In concluding this introduction, we would like to thank S. Adams, D. Gaboriau, J. Melleray, B. Miller, S. Popa, and S. Thomas for providing helpful information or comments related to our work in this paper.

Addendum 1. After the first draft of this paper was completed, we received two recent preprints by N. Monod and Y. Shalom, [MS02,MS02a], which also contain several rigidity results about product groups, in the measure theoretic context, obtained by using the techniques of bounded cohomology. Some of their results seem to have connections to results proved in our paper. For example, in certain cases there is an overlap between Theorem 5.3 below and Theorem 2.22 in the second paper above.

Addendum 2. In connection to the problems discussed in **F)**, Hjorth has recently shown that there are indeed treeable equivalence relations strictly between $E_0, E_{\infty T}$. This also answers negatively the question at the end of **F)**. Also Gaboriau and Popa found continuum many non orbit equivalent measure preserving, ergodic free Borel actions of F_2.

CHAPTER 0

Preliminaries

0A. Actions

Consider an action $(\gamma, x) \in \Gamma \times X \mapsto \gamma \cdot x$ of a group Γ on a set X, sometimes referred to as a Γ-*action* or a Γ-*space*. It induces an equivalence relation on X,

$$xE_\Gamma^X y \Leftrightarrow \exists\gamma \in \Gamma(\gamma \cdot x = y),$$

whose equivalence classes are the orbits of the action. The Γ-*saturation* of $A \subseteq X$ is the set $\Gamma \cdot A = \{\gamma \cdot x : x \in A, \gamma \in \Gamma\}$. If $\Gamma \cdot A = A$, we call A Γ-*invariant*. The *stabilizer* of a point $x \in X$, $\mathrm{Stab}(x)$ or Γ_x, is the subgroup of Γ defined by

$$\Gamma_x = \mathrm{Stab}(x) = \{\gamma \in \Gamma : \gamma \cdot x = x\}.$$

If $\mathrm{Stab}(x) = \{1\}$ for all $x \in X$, i.e., $\gamma \cdot x \neq x, \forall\gamma \in \Gamma, x \in X$, we call the action *free*. In general the *free part* of the action of Γ on X is the Γ-invariant set $\{x \in X : \mathrm{Stab}(x) = \{1\}\})$.

If Γ acts on spaces X, Y, a map $\rho : X \to Y$ is called a Γ-*map* if $\rho(\gamma \cdot x) = \gamma \cdot \rho(x), \forall\gamma \in \Gamma, x \in X$.

For any group Γ and any set X, we have the *shift action* of X on X^Γ, defined by $(\gamma \cdot f)(\delta) = f(\gamma^{-1}\delta)$. The corresponding equivalence relation is denoted by

$$E(\Gamma, X).$$

We let $(X)^\Gamma$ be the free part of this action, i.e., the set

$$(X)^\Gamma = \{f \in X^\Gamma : \gamma \cdot f \neq f, \forall\gamma \in \Gamma, \gamma \neq 1\}.$$

The equivalence relation induced by the action of Γ on $(X)^\Gamma$ is denoted by

$$F(\Gamma, X).$$

If Γ acts on each space $X_i, i \in I$, the *diagonal* action of Γ on $\prod_i X_i$ is given by:

$$\gamma \cdot (x_i) = (\gamma \cdot x_i).$$

If each Γ_i acts on $X_i, i \in I$, then the *product action* of $\prod_i \Gamma_i$ on $\prod_i X_i$ is given by:

$$(\gamma_i) \cdot (x_i) = (\gamma_i \cdot x_i).$$

0B. Equivalence relations

If $E \subseteq X^2$ is an equivalence relation on a set X, we write interchangeably xEy or $(x, y) \in E$ to indicate that x is equivalent to y. We denote by X/E the quotient space, i.e., the set of its equivalence classes. We also denote by $[x]_E$ the equivalence class or E-*class* of $x \in X$. More generally if $A \subseteq X$, we let

$$[A]_E = \{x \in X : \exists y \in A(xEy)\},$$

and call $[A]_E$ the E-*saturation* of A. If $[A]_E = A$, we call A E-*invariant*. If $[A]_E = X$, we call A a *complete section* of E. Finally, $E|A = E \cap A^2$ is the restriction of E to A. A *transversal* for E is a subset $T \subseteq X$ such that T intersects every E-class in exactly one point. We call an equivalence relation E *finite* (resp., *countable*) if all its classes $[x]_E$ are finite (resp., countable). For each set $X, I(X) = X^2$ denotes the coarsest, and $\Delta(X) = \{(x,x) : x \in X\}$ the finest equivalence relation on X. If E, F are equivalence relations on X, then E is a *subequivalence* relation of F, in symbols $E \subseteq F$, if $xEy \Rightarrow xFy, \forall x, y \in X$.

Suppose E_i is an equivalence relation on $X_i, i \in I$. The *product* is the equivalence relation $\prod_i E_i$ on $\prod_i X_i$, given by

$$(x_i) \prod_i E_i (y_i) \Leftrightarrow \forall i \in I (x_i E_i y_i).$$

If E, F are equivalence relations on sets X, Y, resp., a *homomorphism* of E to F is a map $\rho : X \to Y$ such that

$$x_1 E x_2 \Rightarrow \rho(x_1) F \rho(x_2).$$

We call ρ a *reduction* if, moreover,

$$x_1 E x_2 \Leftrightarrow \rho(x_1) F \rho(x_2).$$

A homomorphism ρ induces a map $\bar{\rho} : X/E \to Y/F$, given by $\bar{\rho}([x]_E) = [\rho(x)]_F$. If ρ is a reduction this map is 1-1. Conversely, if $\sigma : X/E \to Y/F$ is a map, a *lifting* of σ is any homomorphism ρ of E to F with $\sigma = \bar{\rho}$.

0C. Borel notions

In this paper we work with standard Borel spaces, i.e., Polish (complete separable metric) spaces equipped with their σ-algebra of Borel sets. An equivalence relation E on X is Borel if it is a Borel subset of X^2.

A Borel *isomorphism* between Borel equivalence relations E, F on standard Borel spaces X, Y, resp., is a Borel bijection $\pi : X \to Y$ which sends E to F, i.e., $x_1 E x_2 \Leftrightarrow \pi(x_1) F \pi(x_2)$. We use

$$E \cong_B F$$

to denote that E, F are Borel isomorphic.

We say that E is *Borel reducible* to F, in symbols

$$E \leq_B F,$$

if there is a Borel reduction of E to F. If there is a 1-1 Borel reduction of E to F, we write $E \sqsubseteq_B F$. We say that E is *Borel bireducible* to F, in symbols,

$$E \sim_B F,$$

if $E \leq_B F$ and $F \leq_B E$. When E, F are countable Borel equivalence relations, the following are equivalent (see [DJK, Prop. 2.6]):

(i) $E \sim_B F$.

(ii) There are Borel sets $A \subseteq X, B \subseteq Y$ which are complete sections for E, F, resp., so that $(E|A) \cong_B (F|B)$.

(iii) There is a "Borel" bijection of X/E onto Y/F, i.e., a bijection $\sigma : X/E \to Y/F$, so that both σ, σ^{-1} admit Borel liftings.

Finally, we let

$$E <_B F \Leftrightarrow E \leq_B F \;\&\; F \not\leq_B E.$$

A Borel equivalence relation E on X is called *tame* (or *smooth*) if there is a Borel map $\rho : X \to Y$, where Y is a standard Borel space, such that

$$x_1 E x_2 \Leftrightarrow \rho(x_1) = \rho(x_2),$$

i.e., $E \leq_B \Delta(Y)$. If E is also countable, this means exactly that E has a Borel transversal. We call E *hypertame* (or *hypersmooth*) if $E = \bigcup_n E_n$, where $E_1 \subseteq E_2 \subseteq \dots$ are tame Borel equivalence relations.

A countable Borel equivalence relation is called *hyperfinite* if it can be written as $E = \bigcup_n E_n$, where $E_1 \subseteq E_2 \subseteq \dots$ are finite Borel equivalence relations. (It can be shown, see [DJK, 5.1], that hyperfinite is equivalent to being countable and hypersmooth.)

Finally, we call a countable Borel equivalence relation E on a standard Borel space X *compressible* if there is a 1-1 Borel map $\rho : X \to X$ such that $xE\rho(x), \forall x$, and for each E-class $C, \rho(C) \subsetneq C$.

Our general references for the descriptive set theory of countable Borel equivalence relations will be [DJK] and [JKL].

0D. Measures

In this paper, measure in a standard Borel space, always means a probability Borel measure on that space.

If μ is a measure on X, a Borel set $A \subseteq X$ is called (μ)-*null* if $\mu(A) = 0$ and (μ)-*conull* if $\mu(A) = 1$.

The *measure class* of a measure μ is the equivalence class of μ under measure equivalence:

$$\begin{aligned} \mu \sim \nu &\Leftrightarrow \mu, \nu \text{ has the same null sets} \\ &\Leftrightarrow \forall \delta \exists \epsilon (\mu(A) < \epsilon \Rightarrow \nu(A) < \delta) \ \& \\ &\quad\ \forall \delta \exists \epsilon (\nu(A) < \epsilon \Rightarrow \mu(A) < \delta), \end{aligned}$$

where A varies over Borel sets.

If μ is a measure on X and $\pi : X \to Y$ is a Borel function, the *image measure* $\pi_* \mu$ is the measure on Y defined by

$$\pi_* \mu(B) = \mu(\pi^{-1}(B)),$$

for every Borel set $B \subseteq Y$.

If μ is a measure on X, we say that E is μ-*hyperfinite* if for some E-invariant conull Borel set A, $E|A$ is hyperfinite.

0E. Borel actions and measures

Let Γ be a countable group and suppose Γ *acts in a Borel way* on a standard Borel space X (i.e., for each $\gamma \in \Gamma, x \mapsto \gamma \cdot x$ is Borel). Then Γ acts (also in a Borel way) on the standard Borel space $\mathcal{M}(X)$ of measures on X (see [Ke95]) by

$$(\gamma \cdot \mu)(A) = \mu(\gamma^{-1} \cdot A),$$

for any Borel set $A \subseteq X$. We say that μ if Γ-*invariant* if $\gamma \cdot \mu = \mu, \forall \gamma \in \Gamma$. We say that μ is Γ-*quasi-invariant* if $\gamma \cdot \mu \sim \mu, \forall \gamma \in \Gamma$ (in that case one also says that the action is *non-singular*).

We generalize this to countable Borel equivalence relations. By a theorem of Feldman-Moore [FM], if E is a countable Borel equivalence relation on X, there is a countable group Γ and a Borel action of Γ on X such that $E_\Gamma^X = E$. We then say that the measure μ is *E-invariant* (resp., *quasi-invariant*) if μ is Γ-invariant (resp., quasi-invariant) for some (equivalently any) countable group Γ and Borel action of Γ on X with $E_\Gamma^X = E$. In the case of quasi-invariance, it is easy to see that this is equivalent to saying that the E-saturation of any null Borel set is null.

Finally, if μ is a measure on X and we have a Borel action of a countable group Γ on X, we say that the measure μ is *Γ-ergodic* or that the Γ-action is *ergodic* (relative to μ) if every Γ-invariant Borel set A is either null or conull. Similarly we say that a countable Borel equivalence relation E on X is *ergodic* (with respect to μ) or that μ is *E-ergodic*, if every E-invariant Borel set is either null or conull.

0F. Amenability

We follow here [JKL, Section 2]. Given a countable set C, a *finitely additive probability measure (f.a.p.)* on C is a map $\varphi : \{A : A \subseteq C\} \to [0,1]$ such that $\varphi(C) = 1, \varphi(A \cup B) = \varphi(A) + \varphi(B)$, if $A \cap B = \emptyset$. A *mean* on C is a positive linear functional $\bar{\varphi}$ on $\ell_\infty(C)$, the Banach space of bounded real functions on C, with $\bar{\varphi}(1) = 1$. Means and f.a.p.'s are the same thing via the identification:

$$\varphi \leftrightarrow \bar{\varphi},$$

where $\bar{\varphi}(f) = \int f d\varphi$ and $\varphi(A) = \bar{\varphi}(1_A)$, with

$$1_A = \text{ the characteristic function of } A.$$

We will not distinguish between $\varphi, \bar{\varphi}$ from now on.

A countable group Γ is *amenable* if there is a left-invariant mean φ on Γ (i.e., a mean φ on Γ such that $\varphi(f) = \varphi(\gamma \cdot f)$, with $\gamma \cdot f(\delta) = f(\delta^{-1}\gamma), \forall \gamma \in \Gamma, f \in \ell_\infty(\Gamma)$).

We also say that a mean φ on $\mathbb{N}$ is *shift-invariant* if $\varphi(f) = \varphi(f_s)$, when $f_s(n) = f(n+1)$ for $f \in \ell_\infty(\mathbb{N})$.

By a result of Christensen [C], Mokobodzki, for each measure μ on $[-1,1]^{\mathbb{N}}$, $\mathbb{N}$ admits a μ-measurable, shift-invariant mean φ (i.e., a shift-invariant mean φ such that $\varphi|[-1,1]^{\mathbb{N}}$ is μ-measurable), and similarly for any amenable group Γ and measure μ on $[-1,1]^\Gamma$ there is left-invariant mean φ on Γ such that $\varphi|[-1,1]^\Gamma$ is μ-measurable. Moreover, assuming the Continuum Hypothesis (CH), φ, in both cases, can be taken to be universally measurable, i.e., μ-measurable for any μ as before. Finally, in all the above, "left-invariant" can be replaced by "right-invariant" or even "(two-sided)-invariant".

A basic result, due to Følner, is that every countable amenable group Γ admits a sequence (F_n) of nonempty finite subsets with the property $\frac{|\gamma F_n \Delta F|}{|F_n|} \to 0, \forall \gamma \in \Gamma$, where $|A| = \text{card } (A)$ is the cardinality of A. Such a sequence is called a *Følner sequence.*

If now E is a countable Borel equivalence relation on a standard Borel space X and μ is a measure on X, we say that E is *μ-amenable* if there is a map $C \mapsto \varphi_C$ assigning to each E-class C a mean φ_C on C, such that if $f : E \to [-1,1]$ is Borel, then $x \mapsto \varphi_{[x]_E}(f_x)$ is μ-measurable, where $f_x(y) = f(x,y)$. We call E *measure amenable* if $x \mapsto \varphi_{[x]_E}(f_x)$ is universally measurable.

If Γ is amenable and acts in a Borel way on a standard Borel space X, then, for any measure μ on X, E_Γ^X is μ-amenable. Since the hyperfinite equivalence relations are exactly those of the form $E_{\mathbb{Z}}^X$, every hyperfinite equivalence relation is μ-amenable for each μ. Finally, by the result of [CFW], μ-amenability is equivalent to

μ-hyperfiniteness. Under CH, measure-amenability is equivalent to (μ-amenability, for every μ) and to (μ-hyperfiniteness, for every μ).

We will also occasionally refer to the concept of α-amenability (for α an ordinal, but we will be only dealing with $\alpha = 1, 2$), for which we refer the reader to [JKL, Section 2.4].

Finally, we will often use the following standard fact (see, e.g., [JKL, 2.5]): If Γ is a countable group which acts freely on a standard Borel space X with invariant measure μ, and if E_Γ^X is μ-amenable, then Γ is amenable.

CHAPTER 1

Actions of Free Groups and Treeable Equivalence Relations

A) First we recall some results from [JKL,§3].

A countable Borel equivalence relation E on a standard Borel space X is *treeable* if it is $\mathcal{T}$-structurable, where $\mathcal{T}$ is the class of trees (connected, acyclic graphs) (see also Appendix D). This means that there is a map $C \in X/E \mapsto T_C$ assigning to each E-class C a tree T_C with vertex set C such that the relation

$$R(x, y, z) \Leftrightarrow (x, y) \in T_{[z]_E}$$

is Borel. This is of course equivalent to saying that there is a Borel acyclic graph with vertex set X, whose connected components are the E-classes.

Treeable equivalence relations are clearly connected to free actions of free groups, in view of the following fact.

Proposition 1.1. *Suppose the free group F_n $(1 \leq n \leq \infty)$ acts freely and in a Borel way on the standard Borel space X. Then $E_{F_n}^X$ is treeable. Conversely, if a countable Borel equivalence relation E is treeable, there is a free Borel action of F_n $(n \geq 2)$ on a standard Borel space X such that*

$$E \sim_B E_{F_n}^X.$$

Proof. The first assertion is obvious. For the second, we can assume that E is compressible (since $E \sim_B E \times I(\mathbb{N})$). By [JKL, 3.1], $E \sqsubseteq_B F(F_2, 2)$, the equivalence relation induced by the shift action of F_2 on $(2)^{F_2}$. But then, by [DJK, 2.3], $E \cong_B F(F_2, 2)|A$, with A a Borel invariant subset of $(2)^{F_2}$, so clearly E is induced by a free Borel action of F_2. ⊣

There is a (unique up to $\sim_B$) largest or *universal* treeable equivalence relation, denoted by $E_{\infty T}$, i.e.,

$$E \leq_B E_{\infty T}$$

for any treeable E. We have

$$E_{\infty T} \sim_B F(F_n, 2),$$

for $2 \leq n \leq \infty$. There is also a (unique up to $\sim_B$) smallest non-tame treeable equivalence relation, denoted by E_0, i.e.,

$$E_0 \leq_B E$$

for any treeable E. Here E_0 can be taken to be the equivalence relation on $2^{\mathbb{N}}$ defined by

$$xE_0y \Leftrightarrow \exists n \forall m \geq n(x_m = y_m),$$

or any hyperfinite non-tame equivalence relation, since these are all the same, up to $\sim_B$, by [DJK, 7.1].

Thus the non-tame treeable equivalence relations are exactly those in the interval

$$E_0 \leq_B E \leq_B E_{\infty T}.$$

Moreover, $E_0 <_B E_{\infty T}$ (see [JKL, 3.5]). However, it is not known whether this interval is non-trivial.

Open Problem 1.2 (see [JKL, 6.4]). Is there a countable Borel equivalence E with

$$E_0 <_B E <_B E_{\infty T}?$$

In fact even the following, much bigger gap in our knowledge, exists.

Open Problem 1.3. Is it true that $E_{\infty T}$ is the smallest non-hyperfinite Borel equivalence relation, i.e.,

$$E_{\infty T} \leq_B E,$$

for any non-hyperfinite countable Borel equivalence relation E?

Addendum. Hjorth has recently provided a positive answer to 1.2, and therefore a negative one to 1.3.

B) Thus we are missing at this stage techniques by which we can distinguish non-hyperfinite treeable equivalence relations, up to $\sim_B$. Noticing that if E is induced by a free Borel action of F_n ($n \geq 2$) admitting an invariant measure, then E is not hyperfinite (see [JKL, 1.7]), we have a rich source of examples of non-hyperfinite treeable equivalence relations. Here is a sample:

(i) $F(F_n, 2)$, the free part of the equivalence relation induced by the shift action of F_n on $2^{F_n}, n \geq 2$.

(ii) Let G be a compact Polish group containing a free subgroup with more than one generator (e.g., $G = \mathrm{SO}(3)$). Thus $F_n \leq G, n \geq 2$. Consider the equivalence relation induced by the left-translation action of F_n on G (whose classes are the right cosets of F_n in G). Since this action leaves the Haar measure of G invariant, it is treeable but not hyperfinite.

(iii) Consider the natural action of $\mathrm{SL}_n(\mathbb{Z})$ on $\mathbb{T}^n$ ($n \geq 2$) (by matrix multiplication). Fix a copy of $F_2 \subseteq \mathrm{SL}_n(\mathbb{Z})$. Then the action of F_2 on $\mathbb{T}^n$ is free a.e. with respect to the standard measure on $\mathbb{T}^n$, which is invariant under this action. So restricted to a co-null set, it gives rise to a treeable non-hyperfinite equivalence relation.

(iv) Similarly consider F_2 as a subgroup of $\mathrm{SO}(n+1)$ ($n \geq 2$) and consider its natural action on S^n.

(v) Let G be the group of Lipschitz automorphisms of $2^{\mathbb{N}}$ (that is the group of automorphisms of the rooted binary tree). It can be seen (see e.g. [SS, Lemma 2]) that there are f, g in G so that every non-trivial reduced word in f, g moves every point of $2^{\mathbb{N}}$. Thus $F_2 = \langle f, g \rangle$ acts freely on $2^{\mathbb{N}}$ and, since it leaves the usual product measure on $2^{\mathbb{N}}$ invariant, it gives rise to a treeable non-hyperfinite equivalence relation.

(vi) Consider a free action of F_n on X_1 with invariant measure μ_1 and *any* action of F_n on X_2 with invariant measure μ_2. The *diagonal action* of F_n on $X_1 \times X_2$ is defined by

$$g \cdot (x_1, x_2) = (g \cdot x_1, g \cdot x_2).$$

It is free with invariant measure $\mu_1 \times \mu_2$. So $E_{F_n}^{X_1 \times X_2}$ is treeable and non-hyperfinite, if $n \geq 2$. Notice that $E_{F_n}^{X_2}$ may well be hyperfinite. This happens, for example, if we take an action of $\mathbb{Z}$ on X_2 with invariant measure μ_2, fix a subjective homomorphism $\varphi : F_n \to \mathbb{Z}$ and let $g \cdot x_2 = \varphi(g) \cdot x_2$ for $g \in F_n$.

C) Although we do not know how to distinguish non-hyperfinite treeable equivalence relations, up to $\sim_B$, there are many important results in ergodic theory that distinguish such relations with respect to stricter notions of equivalence and it is instructive to review them here.

Suppose E_i is a countable Borel equivalence relation on the standard Borel space X_i with invariant measure $\mu_i, i = 1, 2$. We say that $(E_1, \mu_1), (E_2, \mu_2)$ (or just E_1, E_2) are *orbit equivalent* (OE) if there are invariant conull Borel sets $A_i \subseteq X_i, i = 1, 2$, and a Borel isomorphism φ of $E_1|A_1$ with $E_2|A_2$ such that $\varphi_* \mu_1 = \mu_2$. We say that E_1, E_2 are *stably orbit equivalent* (SOE) if there are Borel sets $S_i \subseteq X_i, i = 1, 2$, whose saturations $[S_i]_{E_i}$ are conull, and there is a Borel isomorphism φ of $E_1|S_1$ with $E_2|S_2$ such that $\varphi_*(\mu_1)_{S_1} = (\mu_2)_{S_2}$, where

$$(\mu_i)_{S_i} = \frac{\mu_i|S_i}{\mu_i(S_i)}$$

is the normalized restriction of μ_i to S_i.

Since for countable Borel equivalence relations E_i on $X_i, i = 1, 2, E_1 \sim_B E_2$ iff there are Borel sets $A_i \subseteq X_i$ with $[A_i]_{E_i} = X_i$ such that $E_1|A_1 \cong_B E_2|A_2$ (see [DJK, 2.6]), one can view SOE as a measure theoretic analog of $\sim_B$.

Actions of groups are called *(stably) orbit equivalent* if the corresponding equivalence relations are.

Gaboriau [Ga00] has shown that free measure preserving actions of F_m, F_n are not orbit equivalent if $m \neq n$. This can be also stated as:

Theorem 1.4 (Gaboriau [Ga00]). *Suppose E, F are countable Borel equivalence relations on X, Y resp., each admitting an invariant measure. If E is induced by a free Borel action of F_n and F is induced by a free Borel action of F_m, with $m \neq n$, then $E \ncong_B F$.*

However it is well-known that there are free ergodic measure preserving actions of F_2, F_3 which are SOE (see, e.g., Gaboriau [Ga00, p. 44 or II.15]), so the preceding result fails if OE is replaced by SOE.

We next consider what is known about actions of the *same* free group, say F_2. At this stage only finitely many distinct, up to OE, examples of free measure preserving ergodic actions of F_2 are known. The first property that distinguished such actions was E_0-ergodicity, and this is discussed in Appendix A.7. Very recently Popa [Po] established the existence of a further non-OE example using the theory of von Neumann algebras.

We will conclude this Chapter by reconsidering the method used to provide the first examples of two distinct, up to OE, free measure preserving ergodic actions of F_2, and studying its properties from the descriptive point of view. This method provided an example of a free measure preserving ergodic action of F_2 which is not OE to $F(F_2, 2)$. It would be interesting then to see whether this method could be used to produce examples of E such that $E_0 <_B E <_B E_{\infty T}$.

D) This method is based on example (vi) of **B)**. For a countable group Γ, denote by $(2)^\Gamma$ the free part of the shift action of Γ on 2^Γ:

$$(2)^\Gamma = \{x \in 2^\Gamma : \forall \gamma \neq 1(\gamma \cdot x \neq x)\},$$

where $\gamma \cdot x(\delta) = x(\gamma^{-1}\delta)$. Denote by $E(\Gamma, 2)$ the equivalence relation induced by the shift action of Γ on 2^Γ and by $F(\Gamma, 2)$ its restriction to $(2)^\Gamma$.

Consider now the shift action of F_2 on $(2)^{F_2}$ and let μ be the usual product measure on 2^{F_2}, which of course concentrates on $(2)^{F_2}$. Let also F_2 act in a Borel way on a standard Borel space Y with ergodic, invariant measure ν, so that $E^Y_{F_2}$ is hyperfinite. Consider finally the diagonal (thus free) action of F_2 on $(2)^{F_2} \times Y$, which admits $\mu \times \nu$ as an ergodic, invariant measure. As discussed in the proof of A7.1, $(E^{(2)^{F_2}}_{F_2}, \mu) = (F(F_2, 2), \mu)$ is not OE to $(E^{(2)^{F_2}\times Y}_{F_2}, \mu \times \nu)$. We will discuss here the place of the equivalence relation $E^{(2)^{F_2}\times Y}_{F_2}$, even in the case Y has no invariant measure, in the Borel reducibility hierarchy.

Theorem 1.5. *Let F_2 act in a Borel way on a standard Borel space Y with $E^Y_{F_2}$ hyperfinite. Then exactly one of the following holds:*

(i) $E^{(2)^{F_2}\times Y}_{F_2} \sim_B E_{\infty T}$.

(ii) $E^{(2)^{F_2}\times Y}_{F_2}$ *is an increasing union of a sequence of hyperfinite Borel equivalence relations.*

Proof. We will make use of the following general lemma.

Lemma 1.6. *Suppose Γ is a countable group, X_1, X_2 are two Borel Γ-spaces and π is a Borel Γ-map from X_1 onto X_2, i.e., $\pi(\gamma \cdot x) = \gamma \cdot \pi(x)$. Suppose that $E^{X_2}_\Gamma$ is hyperfinite and all the stabilizers Γ_y, $y \in X_2$, are finitely generated of polynomial growth. Then $E^{X_1}_\Gamma$ is an increasing union of a sequence of hyperfinite Borel equivalence relations.*

Granting this lemma, we will complete the proof.

Since $E^{(2)^{F_2}\times Y}_{F_2}$ is given by a free action of F_2, clearly it is $\leq_B E_{\infty T}$.

Case (1). There is $y_0 \in Y$ such that its stabilizer is a free group isomorphic to some F_n, $2 \leq n \leq \infty$. Call this stabilizer $\Delta \leq F_2$. Let $E^{(2)^{F_2}}_\Delta$ be the subequivalence relation of $E^{(2)^{F_2}}_{F_2}$ given by the restriction of the shift action to Δ. Consider now the shift action of Δ on $(2)^\Delta$. Clearly $E^{(2)^\Delta}_\Delta \sim_B E_{\infty T}$. Now $E^{(2)^\Delta}_\Delta \leq_B E^{(2)^{F_2}}_\Delta$ via the map $p \in (2)^\Delta \mapsto p^* \in (2)^{F_2}$ given by

$$p^*(g) = \begin{cases} p(g), & \text{if } g \in \Delta, \\ 0, & \text{if } g \notin \Delta. \end{cases}$$

We will now show that

$$E^{(2)^{F_2}}_\Delta \leq_B E^{(2)^{F_2}\times Y}_{F_2},$$

so we have $E^{(2)^{F_2}\times Y}_{F_2} \sim_B E_{\infty T}$. To see this consider the map:

$$p \in (2)^{F_2} \mapsto \bar{p} = (p, y_0) \in (2)^{F_2} \times Y.$$

If $\delta \in \Delta$, then $\overline{\delta \cdot p} = (\delta \cdot p, y_0) = (\delta \cdot p, \delta \cdot y_0) = \delta \cdot \bar{p}$. Conversely, if $\bar{p} E_{F_2}^{(2)^{F_2} \times Y} \bar{q}$, then there is $\gamma \in F_2$ with $\gamma \cdot \bar{p} = \bar{q}$ or $\gamma \cdot (p, y_0) = (\gamma \cdot p, \gamma \cdot y_0) = (q, y_0)$, so $\gamma \cdot y_0 = y_0, \gamma \in \Delta$, and $\gamma \cdot p = q$, so $p E_{\Delta}^{(2)^{F_2}} q$.

Case (2). All the stabilizers of the action on Y are isomorphic to $\mathbb{Z}$ or trivial. In that case, noticing that the map $(p, y) \mapsto y$ is an F_2-map from the F_2-space $(2)^{F_2} \times Y$ onto Y, Lemma 1.6 implies that $E_{F_2}^{(2)^{F_2} \times Y}$ is an increasing union of a sequence of hyperfinite Borel equivalence relations.

This completes the proof modulo the lemma.

Proof of Lemma 1.6. Let R be the following equivalence relation on X_1:

$$xRy \Leftrightarrow x E_{\Gamma}^{X_1} x' \ \& \ \pi(x) = \pi(x').$$

Then notice that if for $y \in X_2$, we let

$$X_1(y) = \{x \in X_1 : \pi(x) = y\},$$

then $X_1(y)$ is R-invariant and Γ_y-invariant and $R|X_1(y)$ is induced by the Γ_y-action on $X_1(y)$. Thus it is hyperfinite (see [JKL, 1.20]) uniformly in y. So, as the sets $X_1(y)$ form a partition of X_1, R is hyperfinite.

Now let $\{E_n\}$ be an increasing sequence of finite Borel equivalence relations with $\bigcup_n E_n = E_{\Gamma}^{X_2}$. Let E_n^* be defined on X_1 by:

$$x E_n^* x' \Leftrightarrow x E_{\Gamma}^{X_1} x' \ \& \ \pi(x) E_n \pi(x').$$

Then $\{E_n^*\}$ is increasing and $\bigcup_n E_n^* = E_{\Gamma}^{X_1}$. Moreover, each E_n^*-class contains only finitely many R-classes, so, by [JKL, 1.3], each E_n^* is hyperfinite, thus $E_{\Gamma}^{X_1}$ is an increasing union of a sequence of hyperfinite Borel equivalence relations. $\dashv$

Remarks. (i) Simon Thomas pointed out that if the action of F_2 on Y admits an invariant measure, then (i) of 1.5 holds. This is because in this case F_2 acts freely on $(2)^{F_2} \times Y$ with invariant probability measure, say μ. If (ii) is true, then $E_{F_2}^{(2)^{F_2} \times Y}$, being an increasing union of hyperfinite Borel equivalence relations, is μ-amenable (see, e.g., [JKL, 2.9]), contradicting the fact stated in the last paragraph of 0F.

(ii) The argument for Lemma 1.6 also shows that if all stabilizers Γ_y are *amenable*, then $E_{\Gamma}^{X_1}$ is an increasing union of a sequence of equivalence relations each of which is 1-amenable, thus $E_{\Gamma}^{X_1}$ is 2-amenable (for the definitions, see [JKL, 2.4]). Also if all the stabilizers Γ_y are *finite*, then $E_{\Gamma}^{X_1}$ is actually hyperfinite.

(iii) Note also the following additional observations concerning 1.6:

If X_2 can be split into two Γ-invariant Borel sets Y, Z such that E_{Γ}^{Y} is tame and for $y \in Y$, Γ_y is finitely generated of polynomial growth, while for $z \in Z$, Γ_z is finite, then $E_{\Gamma}^{X_1}$ is also hyperfinite. To see this, note first that $E_{\Gamma}^{X_1}|\pi^{-1}(Z)$ is hyperfinite by the preceding paragraph. Consider now $E_{\Gamma}^{X_1}|\pi^{-1}(Y)$. Let T be a Borel transversal for Y. Then, as before, for any $y \in Y$, $R|\pi^{-1}(\{y\}) = E_{\Gamma}^{X_1}|\pi^{-1}(\{y\})$ is given by a Borel action of Γ_y, uniformly in y, so it is hyperfinite, uniformly in y. It follows that $E_{\Gamma}^{X_1}|\pi^{-1}(T)$ is hyperfinite and, since the Γ-saturation of $\pi^{-1}(T)$ is $\pi^{-1}(Y)$, it follows that $E_{\Gamma}^{X_1}|\pi^{-1}(Y)$ is hyperfinite. Thus $E_{\Gamma}^{X_1}$ is hyperfinite.

For example, if we take $\Gamma = F_2, X_2 =$ the boundary of F_2 (i.e., the space of all infinite reduced words in the generators a, b of F_2) on which F_2 acts by left concatenation and cancelation, then $E_{F_2}^{X_2}$ is hyperfinite and the stabilizers of this action are trivial except on a countable set on which they are isomorphic to $\mathbb{Z}$ (see

Appendix C2). Thus the diagonal action of F_2 on $X_1 \times X_2$, where X_1 is a compact metrizable space on which F_2 acts freely by homeomorphisms (see, e.g., (ii) or (v) of 1B), is an example of a free action of F_2 by homeomorphisms on a compact metrizable space with hyperfinite induced equivalence relation.

In a discussion about whether such an action exists, Adams pointed out that Zimmer [Zi78] proved that an extension of an amenable action (in the measure theoretic context) is also amenable and asked if a similar result is true in the hyperfinite case (in the Borel) context. This motivated the proof of Lemma 1.6.

(iv) It is not known whether an increasing union of a sequence of hyperfinite Borel equivalence relations is hyperfinite (see [JKL, 6.1 (A)]). So those equivalence relations of the form ${E^{(2)}_{F_2}}^{F_2 \times Y}$ which satisfy the hypotheses of Theorem 1.5 and are not hyperfinite, either provide a counterexample to this assertion or else they are the same, up to $\sim_B$, as $E_{\infty T}$.

CHAPTER 2

A Cocycle Reduction Result

We will prove in this section a general cocycle reduction result for E_0-ergodic actions of groups containing infinite amenable normal subgroups, with target hyperbolic groups, and derive various applications in the next Chapter. In order to make the ideas of the proof more transparent, and the treatment of the particular cases, that come up in the applications that we have in mind, as elementary as possible, we will formulate an ad hoc property, which we call near-hyperbolicity, already used, without a name, in [Ad96, Th. 4.8]. This is satisfied by all hyperbolic groups, but is not hard to verify by elementary means for the free groups, and related groups, that come up in the applications.

In the definition below, we use the following notation: For each compact metric space K, we denote by $\mathcal{M}(K)$ the compact, metric space of all measures on K, see, e.g., [Ke95]. Let $\mathcal{M}_{\leq 2}(K)$ be the Borel subset of $\mathcal{M}(K)$ consisting of all $\nu \in \mathcal{M}(K)$ with support of cardinality ≤ 2, i.e., $\nu(\{a, b\}) = 1$ for some $a, b \in K$ (not necessarily distinct). We also let

$$\mathcal{M}_3(K) = \mathcal{M}(K) \setminus \mathcal{M}_{\leq 2}(K).$$

If a countable group H acts continuously on K, it also induces a continuous action on $\mathcal{M}(K)$, defined by

$$h \cdot \nu(A) = \nu(h^{-1} \cdot A),$$

for $A \subseteq K$ Borel (or equivalently for every continuous $f : K \to \mathbb{C}, \int f d(h \cdot \nu) = \int (h^{-1} \cdot f) d\nu$, where $h \cdot f(k) = f(h^{-1} \cdot k)$). Clearly $\mathcal{M}_{\leq 2}(K), \mathcal{M}_3(K)$ are invariant under this action.

Definition 2.1. *A countable group H is* near-hyperbolic *if it admits a continuous action on a compact metric space K with the following properties:*

(a) The induced action of H on $\mathcal{M}_{\leq 2}(K)$ has amenable stabilizers and the corresponding equivalence relation $E_H^{\mathcal{M}_{\leq 2}(K)}$ is λ-hyperfinite for all measures λ on $\mathcal{M}(K)$.

(b) The induced action of H on $\mathcal{M}_3(K)$ has finite stabilizers and the corresponding equivalence relation $E_H^{\mathcal{M}_3(K)}$ is tame.

Thus by Appendix C2.4, F_n, for $1 \leq n \leq \infty$, is near-hyperbolic and by Appendix C4.4, every hyperbolic group is near-hyperbolic. Note also that a subgroup of a near-hyperbolic group is near-hyperbolic.

We now have the following result, where we refer the reader to Appendix B1 for the basic notions concerning cocycles.

Theorem 2.2. *Let Γ be a countable group and $\Delta \trianglelefteq \Gamma$ an infinite normal amenable subgroup. Suppose Γ acts in a Borel way on the standard Borel space X with invariant measure μ. Assume that this action is E_0-ergodic and the action of Δ is* *ergodic. Let H be near-hyperbolic and $\alpha : \Gamma \times X \to H$ a Borel cocycle. Then one of the following holds:*

(i) There is a Borel cocycle $\beta : \Gamma \times X \to H$ equivalent to $\alpha, \alpha \sim \beta$, such that $\beta(\Gamma \times X) \subseteq H_0$, where $H_0 \leq H$ is an amenable subgroup of H.

(ii) There is a Borel cocycle $\beta : \Gamma \times X \to H$ equivalent to $\alpha, \alpha \sim \beta$, such that $\beta(\Delta \times X) \subseteq F_0$, where $F_0 \leq H$ is a finite subgroup of H. And in this case, the double coset

$$\pi(\gamma) = F_0\beta(\gamma, x)F_0, \gamma \in \Gamma,$$

depends only on γ, μ-a.e. (x).

In particular, if H is torsion-free, then (ii) can be replaced by:

(ii)′ There is a homomorphism $\pi : \Gamma \to H$ with $\pi(\Delta) = \{1\}$ such that π (viewed as a cocycle from $\Gamma \times X$ into H via $\pi(\gamma, x) = \pi(\gamma)$) is equivalent to $\alpha, \alpha \sim \pi$.

Proof. Let $\alpha_\Delta = \alpha|\Delta \times X$, so that α_Δ is a Borel cocycle from $\Delta \times X$ into H. Since Δ is amenable, by Appendix B3.1, there is a μ-measurable map

$$x \mapsto \bar{\nu}_x \in \mathcal{M}(K),$$

which is α_Δ-invariant, i.e., for all $\delta \in \Delta$,

$$\alpha(\delta, x) \cdot \bar{\nu}_x = \bar{\nu}_{\delta \cdot x}, \mu\text{-a.e.}(x).$$

We now have two cases:

(i) All μ-measurable maps $x \mapsto \nu_x \in \mu(K)$ which are α_Δ-invariant, are such that $\nu_x \in \mathcal{M}_{\leq 2}(K)$, μ-a.e. (x). Then, by B4.1, there is a maximum such map, which we denote by $x \mapsto \hat{\nu}_x$, and we can assume that $\hat{\nu}_x$ assigns equal mass to the elements of its support. Note that, by Δ-ergodicity, the size of $\text{supp}(\hat{\nu}_x)$ is fixed, μ-a.e. (x). We now check that this map is also α-invariant, i.e., for all $\gamma \in \Gamma$,

$$\alpha(\gamma, x) \cdot \hat{\nu}_x = \hat{\nu}_{\gamma \cdot x}, \mu\text{-a.e.}(x).$$

To see this, fix $\gamma \in \Gamma$ and put

$$\nu'_x = \alpha(\gamma, x)^{-1} \cdot \hat{\nu}_{\gamma \cdot x}.$$

Then $x \mapsto \nu'_x \in \mathcal{M}_{\leq 2}(K)$ is μ-measurable, and we claim that it is α_Δ-invariant, i.e., for $\delta \in \Delta$,

$$\alpha(\delta, x) \cdot \nu'_x = \nu'_{\delta \cdot x} \ \mu\text{-a.e.}(x),$$

or

$$\alpha(\delta, x) \cdot \alpha(\gamma \cdot x)^{-1} \cdot \hat{\nu}_{\gamma \cdot x} = \alpha(\gamma, \delta \cdot x)^{-1} \cdot \hat{\nu}_{\gamma\delta \cdot x}, \ \mu\text{-a.e.}(x).$$

Since $\Delta \trianglelefteq \Gamma$, let $\delta' \in \Delta$ be such that $\gamma\delta = \delta'\gamma$, so that $\hat{\nu}_{\gamma\delta \cdot x} = \hat{\nu}_{\delta'\gamma \cdot x} = \alpha(\delta', \gamma \cdot x) \cdot \hat{\nu}_{\gamma \cdot x}$. Then it is enough to check that, μ-a.e. (x),

$$\alpha(\delta, x)\alpha(\gamma, x)^{-1} = \alpha(\gamma, \delta \cdot x)^{-1}\alpha(\delta', \gamma \cdot x)$$

or, μ-a.e. (x),

$$\alpha(\gamma, \delta \cdot x)\alpha(\delta, x) = \alpha(\delta', \gamma \cdot x)\alpha(\gamma, x).$$

But the left hand side is $\alpha(\gamma\delta, x)$ and the right hand side is $\alpha(\delta'\gamma, x)$, so we are done.

Since $x \mapsto \hat{\nu}_x$ is maximum, we have

$$\nu'_x \leq \hat{\nu}_x, \ \mu\text{-a.e.}(x),$$

in the notation of B4.1. Since the support of ν'_x has the same size as that of $\hat{\nu}_x$ and assigns equal mass to each element of its support, it follows that

$$\nu'_x = \hat{\nu}_x, \ \mu\text{-a.e.}(x),$$

i.e.,

(*) $$\alpha(\gamma, x) \cdot \hat{\nu}_x = \hat{\nu}_{\gamma \cdot x}, \ \mu\text{-a.e.}(x)$$

Define $\rho : X \to \mathcal{M}_{\leq 2}(K)$ by

$$\rho(x) = \hat{\nu}_x.$$

Let $\lambda = \rho_* \mu$, a measure on $\mathcal{M}_{\leq 2}(K)$. Since $E_H^{\mathcal{M}_{\leq 2}(K)}$ is λ-hyperfinite, fix an H-invariant Borel subset $Y_0 \subseteq \mathcal{M}_{\leq 2}(K)$ and a Γ-invariant Borel subset $X_0 \subseteq X$ with $\mu(X_0) = 1, \rho(X_0) \subseteq Y_0, E_H^{Y_0}$ hyperfinite, and (by $(*)$) such that $\rho : X_0 \to Y_0$ is a homomorphism of $E_\Gamma^{X_0}$ to $E_H^{Y_0}$.

Since $E_\Gamma^{X_0}$ is E_0-ergodic, it follows, by B2.1, that there is some fixed $\nu_0 \in Y_0$ and $\beta \sim \alpha$ such that $\beta(\Gamma \times X) \subseteq H_0 =$ stabilizer of ν_0, which is amenable, by condition (a) of near-hyperbolicity.

(ii) There is a μ-measurable map $x \mapsto \nu_x \in \mathcal{M}(K)$ which is α_Δ-invariant but $\nu_x \in \mathcal{M}_3(K)$ on a set of μ-positive measure. Then since, μ-a.e. (x), $\nu_x \in \mathcal{M}_3(K)$ iff $\alpha(\delta, x) \cdot \nu_x = \nu_{\delta \cdot x} \in \mathcal{M}_3(K)$, for any $\delta \in \Delta$, it follows, by the ergodicity of the Δ-action, that $\nu_x \in \mathcal{M}_3(K)$, μ-a.e. (x). Then, on a conull set, $x \mapsto \nu_x$ is a homomorphism from E_Δ^X to $E_H^{\mathcal{M}_3(K)}$, which is tame, so by condition (b) of near-hyperbolicity and the ergodicity of the Δ-action again, we conclude, using B2.1 again, that there is a Borel cocycle β, with $\alpha \sim \beta$ and $\beta(\Delta \times X) \subseteq F_0$, where $F_0 \leq H$ is the stabilizer of some $\nu_0 \in \mathcal{M}_3(K)$, thus it is finite.

We verify in this case that also $\pi(\gamma) = F_0 \beta(\gamma, x) F_0, \gamma \in \Gamma$, depends only on γ, μ-a.e. (x). To see this, fix γ and let

$$\theta(x) = F_0 \beta(\gamma, x) F_0.$$

It is enough to show, by the ergodicity of the Δ-action, that

$$\theta(\delta \cdot x) = \theta(x), \ \mu\text{-a.e. } (x).$$

Fix $\delta' \in \Delta$ with $\gamma\delta = \delta'\gamma$. Then $\theta(\delta \cdot x) = F_0 \beta(\gamma, \delta \cdot x) F_0 = F_0 \beta(\gamma\delta, x) \beta(\delta, x)^{-1} F_0 = F_0 \beta(\gamma\delta, x) F_0 = F_0 \beta(\delta'\gamma, x) F_0 = F_0 \beta(\delta', \gamma \cdot x) \beta(\gamma, x) F_0 = F_0 \beta(\gamma, x) F_0 = \theta(x)$, μ-a.e. (x), since $\beta(\Delta \times X) \subseteq F_0$.

Finally, note that if H is torsion-free, $F_0 = \{1\}$, so $\beta(\Delta \times X) = \{1\}$ and

$$\pi(\gamma) = \beta(\gamma, x) \ \mu\text{-a.e.}(x),$$

thus

$$\begin{aligned} \pi(\gamma_1 \gamma_2) &= \beta(\gamma_1 \gamma_2, x) \\ &= \beta(\gamma_1, \gamma_2 \cdot x) \beta(\gamma_2, x), \ \mu\text{-a.e. } (x) \\ &= \pi(\gamma_1) \pi(\gamma_2), \end{aligned}$$

so π is a homomorphism with $\pi(\Delta) = \{1\}$. ⊣

Remark. Theorem 2.2 is related to Theorem 5.4 in Adams [Ad95], with the new ingredient being the use of E_0-ergodicity to refine the analysis and derive stronger conclusions in our case.

CHAPTER 3

Some Applications

We will now use Theorem 2.2 to derive various applications concerning Borel reducibility.

3A. An "elementary" proof of existence of incomparables

We will actually find an infinite family of countable groups $\{\Gamma_p\}$ and a Borel free action of each Γ_p on a standard Borel space X_p, with invariant measure μ_p, so that if $p \neq q$, $E_{\Gamma_p}^{X_p}$ is $E_{\Gamma_q}^{X_q}$-ergodic. In particular, $E_{\Gamma_p}^{X_p} \not\leq_B E_{\Gamma_q}^{X_q}$, if $p \neq q$.

For prime $p \geq 3$, let

$$\Gamma_p = (\mathbb{Z}_p * \mathbb{Z}_p) \times \mathbb{Z},$$

where $\mathbb{Z}_p = \mathbb{Z}/p\mathbb{Z}$, and let $X_p = (2)^{\Gamma_p}$ = the free part of the shift action of Γ_p on 2^{Γ_p}. Let $E_p = F(\Gamma_p, 2)$ be the associated equivalence relation and μ_p the usual product measure on X_p.

Theorem 3.1. *If p, q are distinct odd primes, then E_p is E_q-ergodic. In particular,*

$$E_p \not\leq_B E_q.$$

Proof. The group $\mathbb{Z}_q * \mathbb{Z}_q$ is of course hyperbolic. In Appendix C3 we actually give an explicit elementary proof that $\mathbb{Z}_q * \mathbb{Z}_q$ is near-hyperbolic, and this is all that we are going to use here.

Now assume that

$$\rho : X_p \to X_q$$

is a Borel homomorphism of E_p to E_q. Since Γ_q acts freely on X_q, let α be the associated cocycle $\alpha : \Gamma_p \times X_p \to \Gamma_q$, given by

$$\alpha(\gamma, x) \cdot \rho(x) = \rho(\gamma \cdot x),$$

for $\gamma \in \Gamma_p$, $x \in X_p$. Write

$$\alpha = (\alpha_1, \alpha_2),$$

where

$$\alpha_1 = p_1 \circ \alpha,$$

$$\alpha_2 = p_2 \circ \alpha,$$

with $p_1 : \Gamma_q \to (\mathbb{Z}_q * \mathbb{Z}_q), p_2 : \Gamma_q \to \mathbb{Z}$ the two projections. So α_1, α_2 are cocycles of the Γ_p-action on X_p to $H = \mathbb{Z}_q * \mathbb{Z}_q, \mathbb{Z}$, resp.

We now apply Theorem 2.2 to $\Gamma = \Gamma_p, \Delta = \mathbb{Z}, X = X_p, \mu = \mu_p, \alpha_1, H$. The hypotheses about (Γ, Δ, X, μ) are satisfied by Appendices A4.1, A6.1, and about H by the above remarks. So we have that one of (i), (ii) of 2.2 is true.

If (i) holds, then $\alpha_1 \sim \beta_1$, say via the Borel function $x \mapsto h_x \in H$,

$$\alpha_1(\gamma, x) = h_{\gamma \cdot x} \beta_1(\gamma, x) h_x^{-1}, \ \mu\text{-a.e. } (x),$$

where $\beta_1(\Gamma \times X) \subseteq H_0 \leq H$, an amenable subgroup of H. Now it is well known that an amenable subgroup of H is cyclic. (This follows immediately from Kurosh's Theorem, see [Ro, 11.55]. It is easier to see that H_0 is cyclic-by-finite, using the standard fact that H is free-by-finite, see, e.g., [dlH, IIA.16], or [LS, p. 177], and this is enough for our purposes.) Also

$$\alpha \sim (\beta_1, \alpha_2),$$

via the Borel map $x \mapsto f_x = (h_x, 1)$, therefore

$$\alpha \sim \beta$$

for the Borel cocycle $\beta = (\beta_1, \alpha_2)$, with $\beta(\Gamma \times X) \subseteq \Gamma_0 \subseteq \Gamma_q$, where $\Gamma_0 = H_0 \times \mathbb{Z}$. Put

$$\sigma(x) = f_x^{-1} \cdot \rho(x).$$

Then, since $\alpha(\gamma, x) = f_{\gamma \cdot x}\beta(\gamma, x)f_x^{-1}$, μ-a.e. (x), σ is a Borel homomorphism of E_p into E_q with associated cocycle $\beta(\gamma, x)$, μ-a.e. (x). It follows that if F is the equivalence relation $E_{\Gamma_0}^{X_q}$ induced by the Γ_0-action on X_q, then, on a co-null set in X, σ is a Borel homomorphism of E_p into F. But F is hyperfinite, so σ maps into a single F-class μ-a.e. (x), and thus $\rho(x) = f_x \cdot \sigma(x)$ maps into a single E_q-class a.e., which completes the proof in case (i).

Now suppose that we are in case (ii). Then $\alpha_1 \sim \beta_1$, where $\beta_1(\Delta \times X) \subseteq F_0, F_0 \leq H$ a finite subgroup of H. Moreover $\pi(\gamma) = F_0\beta_1(\gamma, x)F_0$, for $\gamma \in \Gamma$, depends only on γ, μ-a.e. (x). Since F_0 is a finite subgroup of $H = \mathbb{Z}_q * \mathbb{Z}_q = A * B$, where we write A for the first copy of $\mathbb{Z}_q$ and B for the second, it follows that there is $h_0 \in H$ with $h_0F_0h_0^{-1} \subseteq A$ or $h_0F_0h_0^{-1} \subseteq B$ (see, e.g. [Ro, 11.57]). Without loss of generality, we assume that $h_0F_0h_0^{-1} \subseteq A$. Let $\bar{\beta}_1(\gamma, x) = h_0\beta_1(\gamma, x)h_0^{-1}$. Then $\bar{\beta}_1 \sim \beta_1$ and $\bar{\beta}_1(\Delta \times X) \subseteq A$. So replacing β_1 by $\bar{\beta}_1$, if necessary, we can assume, to start with, that already $\beta_1(\Delta \times X) \subseteq A$ and $\pi(\gamma) = A\beta_1(\gamma, x)A$ depends only on γ, μ-a.e. (x).

Let now γ_0 be an element of order p in $\mathbb{Z}_p * \mathbb{Z}_p$. Then $\gamma_0^p = 1$, so

(*) $$\beta_1(\gamma_0^p, x) = \beta_1(\gamma_0, \gamma_0^{p-1} \cdot x) \dots \beta_1(\gamma_0, \gamma_0 \cdot x)\beta_1(\gamma_0, x) = 1.$$

Fix $h \in H$ so that

(**) $$AhA = A\beta_1(\gamma_0, x)A, \ \mu\text{-a.e. } (x)$$

Then, from $(*), (**)$, we have that there are $\bar{a}_1, \dots, \bar{a}_{p+1}$ in A such that

$$\bar{a}_1 h \bar{a}_2 h \bar{a}_3 \dots \bar{a}_p h \bar{a}_{p+1} = 1.$$

Claim. $h \in A$.

Granting this claim, we conclude that

$$\beta_1(\gamma_0, x) \in A, \ \mu\text{-a.e. } (x).$$

Applying this to the generators of $\mathbb{Z}_p * \mathbb{Z}_p$, we conclude that for all $\gamma \in \mathbb{Z}_p * \mathbb{Z}_p$,

$$\beta_1(\gamma, x) \in A, \ \mu\text{-a.e. } (x).$$

It follows that $\alpha \sim \beta$, where $\beta = (\beta_1, \alpha_2)$, and for all $\gamma \in \mathbb{Z}_p * \mathbb{Z}_p$,

$$\beta(\gamma, x) \in A \times \mathbb{Z}, \ \mu\text{-a.e. } (x).$$

Suppose $x \mapsto p_x \in H \times \mathbb{Z}$ is Borel with $\alpha(\gamma, x) = p_{\gamma \cdot x}\beta(\gamma, x)p_x^{-1}$, μ-a.e. (x), for $\gamma \in \Gamma$, and put $\tau(x) = p_x^{-1} \cdot \rho(x)$. Then τ is a Borel homomorphism of E_p into E_q with associated cocycle $\beta(\gamma, x)$, μ-a.e. (x). But as $\beta(\gamma, x)$ takes values in

$A\times\mathbb{Z}$, μ-a.e. (x), for $\gamma\in\mathbb{Z}_p*\mathbb{Z}_p$, it follows that τ is a homomorphism from $E^X_{\mathbb{Z}_p*\mathbb{Z}_p}$, restricted to a conull set, to $E^{X_q}_{A\times\mathbb{Z}}$, which is hyperfinite. Since $E^X_{\mathbb{Z}_p*\mathbb{Z}_p}$ is E_0-ergodic, by A4.1, it follows that τ maps into a single $E^{X_q}_{A\times\mathbb{Z}}$-class, μ-a.e., so ρ maps into a single E_q-class, μ-a.e., and the proof is complete.

So it only remains to prove the claim. This follows from the following algebraic lemma and the fact that A has no non-trivial elements of order p.

Lemma 3.2. *Suppose A,B are groups, $n\geq 1, h=a_1b_1a_2b_2\dots a_nb_n\in A*B$, with $a_i\in A, b_i\in B, b_1,a_2,b_2,\dots,a_n,b_n$ all different from 1, and $\bar{a}_1h\bar{a}_2h\bar{a}_3\dots\bar{a}_ph\bar{a}_{p+1}=1$ with $\bar{a}_i\in A, p\geq 2$. Assume also that A,B have no involutions. If c is the middle element in the sequence $b_1,a_2,b_2,\dots,a_n,b_n$, then $c^p=1$.*

Proof. In case $n=1, h=a_1b_1$, so that $a'_1b_1a'_2b_1\dots a'_pb_1a'_{p+1}=1$, for some $a'_i\in A$. Since $b_1\neq 1$, it is clear that $b_1^p=1$.

Assume now $n\geq 2$. We have $a'_1(b_1a_2\dots a_nb_n)a'_2(b_1a_2\dots a_nb_n)a'_3\dots a'_p(b_1a_2\dots a_nb_n)a'_{p+1}=1$, for some $a'_i\in A$. Then there is $1<i<p+1$ with $a'_i=1$, and we have a largest number of cancelations

$$b_nb_1=1, a_na_2=1,\dots,$$

but these cannot reach the middle element c of the sequence $b_1,a_2,\dots a_n,b_n$, otherwise $c^2=1$. It follows that all $a'_2,\dots,a'_p$ are 1 and $b_nb_1=1, a_na_2=1,\dots$ are all canceled up to the middle element c of the sequence

$$b_1,a_2,\dots,a_nb_n,$$

so that $c^p=1$. ⊣

Remark. J. Melleray pointed out that this lemma is true even without assuming that A,B have no involutions.

3B. Further "elementary" proofs of theorems of Adams-Kechris

As can be seen by reading sections 4, 5 in [AK], to derive Theorems 4.1 and 5.1 in that paper, one only needs to provide examples that satisfy Lemma 4.5 in [AK].

For each non-empty subset S of odd primes, let

$$\Gamma_S=(*_{p\in S}(\mathbb{Z}_p*\mathbb{Z}_p))\times\mathbb{Z}.$$

Let also $X_S=(2)^{\Gamma_S}, E_S=F(\Gamma_S,2)$. Then we have the following result, which provides the desired examples.

Theorem 3.3. *If $S\not\subseteq T$, then E_S is E_T-ergodic.*

Proof. Fix an odd prime $p\in S\setminus T$. Fix a Borel homomorphism

$$\rho: X_S\to X_T$$

of E_S into E_T. Let α be the associated cocycle, and let $\alpha=(\alpha_1,\alpha_2)$, where $\alpha_1=p_1\circ\alpha, \alpha_2=p_2\circ\alpha, p_1,p_2$ the two projections of Γ_T on $*_{q\in T}(\mathbb{Z}_q*\mathbb{Z}_q),\mathbb{Z}$ resp.

Now consider the subgroup $\Gamma_p=(\mathbb{Z}_p*\mathbb{Z}_p)\times\mathbb{Z}$ and its action on X_S and apply Theorem 2.2 to $\Gamma=\Gamma_p, \Delta=\mathbb{Z}, X=X_S, \mu=\mu_S$ (= the usual product measure on $(2)^{\Gamma_S}$), $\alpha_1, H=*_{q\in T}(\mathbb{Z}_q*\mathbb{Z}_q)$. The hypotheses about (Γ,Δ,X,μ) are satisfied by Appendices A4.1, A6.1, and about H by Appendix C3.3. So one of the alternatives (i), (ii) of 2.2 hold. Now (i) can be handled exactly as in the proof of 3.1 (we again have that an amenable subgroup of H is cyclic, by Kurosh's Theorem, see, [Ro, 11.55]).

Now consider case (ii), so that $\alpha_1 \sim \beta_1$, where $\beta_1(\Delta \times X) \subseteq F_0, F_0 \leq H$ a finite subgroup of H and $\pi(\gamma) = F_0\beta_1(\gamma, x)F_0$, for $\gamma \in \Gamma$, depends only on γ, μ-a.e. (x). Again F_0 is conjugate to a subgroup of some $\mathbb{Z}_q, q \in T$, which we can assume it is the first copy in $\mathbb{Z}_q * \mathbb{Z}_q$. So by changing β_1 a bit if necessary, as in the proof of 3.1, we can assume that $\beta_1(\Delta \times X) \subseteq A = \mathbb{Z}_q$ and $\pi(\gamma) = A\beta_1(\gamma, x)A$ depends only on γ, μ-a.e. (x). Again as in the proof of 3.1, we will complete the proof if we can show that $\beta_1(\gamma, x) \in A$, μ-a.e. (x), and for that it is enough to show that if γ_0 is an element of order p in $\mathbb{Z}_p * \mathbb{Z}_p$ and $h \in H$ is such that

$$AhA = A\beta_1(\gamma_0, x)A \ \mu\text{-a.e. } (x),$$

then $h \in A$. Again we have $\bar{a}_1, \dots, \bar{a}_{p+1}$ in A so that $\bar{a}_1 h \bar{a}_2 h \bar{a}_3 \dots \bar{a}_p h \bar{a}_{p+1} = 1$. Suppose $h \notin A$, towards a contradiction. We can assume that $h \in A * B$, where B is a subgroup of H of the form $B = \mathbb{Z}_{q_1} * \mathbb{Z}_{q_2} * \cdots * \mathbb{Z}_{q_n}$ for $q_1, \dots, q_n \in T$, and $h = a_1 b_1 a_2 b_2 \dots a_n b_n$, with $a_i \in A, b_i \in B, n \geq 1$ and all $b_1, a_2, \dots, b_n \neq 1$. Then by Lemma 3.2, there will be some $c \in (A \cup B) \setminus \{1\}$ with $c^p = 1$. But c, being of finite order, is a conjugate of some element of one of $\mathbb{Z}_q, \mathbb{Z}_{q_1}, \dots, \mathbb{Z}_{q_n}$, so $c^p \neq 1$, as $p \notin \{q, q_1, \dots, q_n\}$, a contradiction. ⊣

3C. "Elementary" proofs of results of Adams and Thomas

Using Zimmer's superrigidity theory as well as Ratner's measure classification theorem [Ra], Adams [Ad02] constructed the first examples of countable Borel equivalence relations E, F, on an uncountable Polish space, with $E \subseteq F$ but $E \not\leq_B F$. Thomas [T02a] then used these techniques to solve two other well-known problems, by constructing the first example of a countable Borel equivalence relation E, on an uncountable Polish space, satisfying $E <_B 2E$ and also the first examples of aperiodic countable Borel equivalence relations E, F such that $E \sim_B F$ but for which $E \approx_B F$ fails. (Here $E \approx_B F$ means that $E \sqsubseteq_B F$ and $F \sqsubseteq_B E$, where $E \sqsubseteq_B F$ signifies that there is an *injective* Borel reduction of E to F.) We provide here alternative proofs of these results, that avoid the use of superrigidity and Ratner's theorem.

We will first need the following corollaries of 2.2.

Lemma 3.4. *Suppose $\Gamma = F_2 \times F_2$ acts freely in a Borel way on the standard Borel space X with invariant measure μ, so that the action of each non-amenable subgroup is E_0-ergodic and the action of each infinite subgroup is ergodic. Let H_2 be a countable non-amenable, torsion-free, hyperbolic group and let $H = F_2 \times H_2$. Let $\alpha : \Gamma \times X \to H$ be a Borel cocycle such that its restriction to any non-amenable subgroup is not equivalent to a cocycle taking values in an amenable subgroup of H. Then, if $F_2 = \langle a, b \rangle$, $\alpha | (\langle a \rangle \times \{1\}) \times X \to H$ is equivalent to a non-trivial homomorphism $\pi : \langle a \rangle \times \{1\} \to H$.*

Proof. Let $\alpha_{11} : (\langle a \rangle \times F_2) \times X \to F_2, \alpha_{12} : (\langle a \rangle \times F_2) \times X \to H_2, \alpha_{21} : (F_2 \times \langle a \rangle) \times X \to F_2, \alpha_{22} : (F_2 \times \langle a \rangle) \times X \to H_2$ be the projections of α, restricted to the appropriate subgroups, to the first and second factors of H. Our assumptions and 2.2 imply that one of α_{11}, α_{12} is equivalent to a homomorphism, which is trivial on $\langle a \rangle \times \{1\}$, and has non-amenable range. Similarly for one of α_{21}, α_{22}. Assume without loss of generality that α_{11} has this property. We claim then that α_{22} has also this property, and thus $\alpha | (\langle a \rangle \times \{1\}) \times X$ is equivalent to a non-trivial homomorphism. Indeed, otherwise α_{21} does. Then it follows that the projection to F_2 of

$\alpha|(\langle a\rangle \times \{1\}) \times X$ is equivalent to both a trivial and a non-trivial homomorphism. This contradicts the following fact.

Sublemma 3.5. *Let $\mathbb{Z}$ act freely in a Borel way on a standard Borel space X, with invariant measure μ. Let $\beta : \mathbb{Z} \times X \to G$ be a Borel cocycle into a countable torsion-free group G. Then β cannot be equivalent to both a trivial and a non-trivial homomorphism.*

Proof. Otherwise there is a non-trivial homomorphism $\pi : \mathbb{Z} \to G$ which is equivalent to the trivial homomorphism, i.e., there is a Borel function $f : X \to G$ such that $\forall n \in \mathbb{Z}$

$$\pi(n) = f(n \cdot x) f(x)^{-1}, \ \mu\text{-a.e.}(x),$$

or

$$f(n \cdot x) = \pi(n) f(x), \ \mu\text{-a.e.}(x).$$

Say $\pi(1) = a$, so that $\pi(\mathbb{Z}) = \langle a\rangle$. Choose a subset $T \subseteq G$ meeting every right coset $\langle a\rangle g, g \in G$ in exactly one point. Since $f|(\mathbb{Z} \cdot x)$ is a bijection between $\mathbb{Z} \cdot x$ and $\langle a\rangle f(x)$, let $g(x) =$ the element $y \in \mathbb{Z} \cdot x$ such that $f(y) \in T$. Then $g : X \to X$ is a Borel selector for $E_{\mathbb{Z}}^X$, i.e., $E_{\mathbb{Z}}^X$ is tame, contradicting the fact that $\mathbb{Z}$ acts freely with invariant measure. ⊣

Lemma 3.6. *Let H_2 be a countable non-amenable group which cannot be mapped homomorphically onto a non-abelian free group. Let $H = F_2 \times H_2$ act freely in a Borel way on a standard Borel space Y with invariant measure ν and assume that every non-amenable subgroup of H acts E_0-ergodically and every infinite subgroup acts ergodically. Let $\Gamma = F_2 \times F_2$ act freely in a Borel way on a standard Borel space X. Then $E_H^Y \not\leq_B E_\Gamma^X$.*

Proof. Let $\rho : Y \to X$ be a Borel reduction of E_H^Y to E_Γ^X, towards a contradiction. let α be the associated cocycle, and let α_1, α_2 be its projections to the first and second copy of F_2 in Γ. By 2.2, and our assumptions, $\alpha_i|(\langle a\rangle \times H_2) \times Y$ will be equivalent to a homomorphism, trivial on $\langle a\rangle \times \{1\}$, with non-amenable range, for some $i \in \{1, 2\}$, where $F_2 = \langle a, b\rangle$. Say it is $\alpha_1|(\langle a\rangle \times H_2) \times Y$. Then this gives a homomorphism from H_2 to F_2 with free non-abelian range, a contradiction. ⊣

We will finally need the concept of unique ergodicity. (The idea of using unique ergodicity in these problems, as in the proof of 3.8 below, goes back to Adams [Ad02].) A Borel action of a countable group Γ on a standard Borel space X is *uniquely ergodic* if there is exactly one Γ-invariant measure on X (which must then necessarily be Γ-ergodic). It is a standard fact (e.g., a consequence of the ergodic decomposition theorem, see Farrell [F], Varadarajan [V]) that if the action of Γ on X has an invariant ergodic measure μ, then there is a Γ-invariant Borel set $X_0 \subseteq X$, with $\mu(X_0) = 1$, such that the Γ-action on X_0 is uniquely ergodic (with corresponding measure μ restricted to X_0). (In case $\Gamma = \mathbb{Z}$, which is really all we will use below, this can be easily proved using the Birkhoff ergodic theorem: Fix a countable Boolean algebra $\{B_k\}$ of Borel subsets of X which generate the Borel sets. Let T be the Borel automorphism of X that gives the $\mathbb{Z}$-action. By the ergodic theorem,

$$\frac{\text{card}(\{i \in [-n, n] : T^i(x) \in B_k\})}{2n+1} \to \mu(B_k),$$

for every k, μ-a.e. (x). Define X_0 to be the set of x's for which this happens for all k.)

We thus have:

Lemma 3.7. *Let Γ be a countable group acting in a Borel way on a standard Borel space X with invariant measure μ. If the action is mixing, then there is a Γ-invariant Borel set $X_0 \subseteq X$ such that $\mu(X_0) = 1$, and every infinite cyclic subgroup of Γ acts uniquely ergodically on X_0.*

We are now ready to give the new proof of

Theorem 3.8 (Adams [Ad02]). *There are countable Borel equivalence relations E, F with $E \subseteq F$ and $E \not\leq_B F$.*

Proof. We fix a countable torsion-free, hyperbolic group H_2, containing a copy of F_2, which cannot be mapped homomorphically onto a non-abelian free group. (We will discuss specific examples of such H_2 below.) Let $H = F_2 \times H_2$. Since H_2 contains a copy of F_2, we can view $\Gamma = F_2 \times F_2$ as a subgroup of H. Consider the free part $(2)^H$ of the shift action of H on 2^H and the usual product measure μ. We recall from A6.1 that this action is mixing, so every infinite subgroup acts ergodically, and from A4.1 that every non-amenable subgroup acts E_0-ergodically. Fix an H-invariant Borel subset X of $(2)^H$ such that $\mu(X) = 1$ and every cyclic subgroup of H acts uniquely ergodically on X. We also write μ for the restriction of μ to X.

We now let

$$E = E_\Gamma^X, F = E_H^X.$$

Clearly $E \subseteq F$. So it is enough to check that $E \not\leq_B F$. Suppose $\rho : X \to X$ is a Borel reduction of E_Γ^X to E_H^X and let $\alpha : \Gamma \times X \to H$ be the corresponding cocycle. Then, by 3.4, $\alpha|(\langle a \rangle \times \{1\}) \times X \sim \pi$, where $\pi : \langle a \rangle \times \{1\} \to H$ is a non-trivial homomorphism. As usual, we can then modify ρ to another Borel reduction σ with $\rho(x) E_H^X \sigma(x), \mu$-a.e. (x), so that if β is the cocycle associated to σ, then $\beta(\gamma, x) = \pi(\gamma), \forall \gamma \in \langle a \rangle \times \{1\}, \mu$-a.e. (x) (see B.1). Thus $\sigma(\gamma \cdot x) = \pi(\gamma) \cdot \sigma(x), \forall \gamma \in \langle a \rangle \times \{1\}), \mu$-a.e. (x). Let $\nu = \sigma_* \mu$ be the image of μ under σ.

Claim. If $\pi(\langle a \rangle \times \{1\}) = Z \subseteq H$, then ν is Z-invariant.

Proof. Fix a Borel set $A \subseteq X$. Then for $\delta = \pi(\gamma) \in Z$, where $\gamma \in \langle a \rangle \times \{1\}$, we have $\sigma^{-1}(\delta \cdot A) = \gamma \cdot \sigma^{-1}(A)$, so $\nu(\delta \cdot A) = \mu(\sigma^{-1}(\delta \cdot A)) = \mu(\gamma \cdot \sigma^{-1}(A)) = \mu(\sigma^{-1}(A)) = \nu(A)$.

Then by unique ergodicity (since Z is cyclic) $\mu = \nu$. It follows that $\mu(\sigma(X)) = 1$, and, by using a Borel inverse to σ, we see that there is a Borel conull H-invariant set $Y \subseteq X$ with $E_H^Y \leq_B E_\Gamma^X$. This contradicts 3.6.

We finally comment on the existence of the group H_2. Here is an example provided to us by S. Thomas: Let $H_2 = \langle a, b | aba^2ba^3b \dots ba^n \rangle$. If n is large enough, H_2 satisfies the small cancelation property $C'(1/6)$ (see [LS, p. 240] or [GdlH, p. 227]). We take $n = 22$, which is enough. Thus H_2 is hyperbolic (see [GdlH, p. 254]). It is torsion-free, since the word $aba^2ba^3b \dots ba^{22}$ is not a proper power (see [LS, II, 5.18]). Also H_2 contains a copy of F_2. For example, $a^{23}b^{23}, a^{46}b^{46}$ generate a free group, since any nonempty freely reduced word in a, b, generated by these two elements, cannot contain more than half of an initial segment of a cyclic permutation of the relator of H_2 or its inverse. Thus, by Dehn's algorithm (see [GdlH, p. 244]), it cannot represent the identity of H_2. Finally, H_2 cannot be mapped homomorphically onto a free non-abelian group F, since then the image of the two generators a, b would be a basis for F (see, e.g., [LS, I, 2.7]), which is obviously wrong. $\dashv$

The equivalence relation E used in the proof of 3.8 can be also recycled to prove the results of Thomas [T02a].

Theorem 3.9 (Thomas [T02a]). *There is a countable Borel equivalence relation E with $nE <_B (n+1)E, \forall n \geq 1$, and $E \times I(2) \not\leq_B E$.*

Here nE is the (disjoint) sum of n copies of E, i.e. the equivalence relation on $X \times n$ (where E lives on X), given by $(x,i)nE(y,j) \Leftrightarrow i = j \;\&\; xEy$ and $E \times I(2)$ is the equivalence relation on $X \times 2$ given by

$$(x,i)E \times I(2)(y,j) \Leftrightarrow xEy.$$

Thus clearly $E \times I(2) \sim_B E$.

Proof of 3.9. We let $E = E_\Gamma^X$ be as in the proof of 3.8. If $(n+1)E <_B nE$ for some $n \geq 1$, then, by ergodicity of μ and the pigeonhole principle, we see that there is a Borel Γ-invariant set $Y \subseteq X$, with $\mu(Y) = 1$ and two reductions $\rho_1 : Y \to X, \rho_2 : Y \to X$ of E_Γ^Y to E_Γ^X with $\rho_1(Y) \cap \rho_2(Y) = \emptyset$. As in the proof of 3.8 (using 3.4 again but with $H_2 = F_2$ this time), we can assume that $(\rho_1)_*\mu = (\rho_2)_*\mu = \mu$, so that $\rho_1(Y) \cap \rho_2(Y) \neq \emptyset$, a contradiction.

The proof of $E \times I(2) \not\leq_B E$ is similar. $\dashv$

At the expense of making the argument more technical, one can also give a proof of 3.8 (and 3.9) that uses the group $\mathbb{Z}_p * \mathbb{Z}$ instead of H_2 as defined above, thus making the proof even more "elementary". Since the ideas involved can probably be used to deal with other problems, in this context, concerning groups that have torsion, we present this proof below.

Let Γ be a group acting on a probability space (X, μ). A cocycle $\beta : \Gamma \times X \to G$ is *diffuse* if the function

$$\gamma \mapsto \beta(\gamma, x)$$

is finite-to-one a.e; the cocycle is *narrow* if

$$\gamma \mapsto \beta(\gamma, x)$$

has finite image a.e.

Clearly when Γ is infinite these are opposing concepts. We observe in the next lemma that these are opposing even up to equivalence of cocycles.

Lemma 3.10. *Let Γ be a countably infinite group acting freely by measure preserving transformations on a standard Borel probability space (X, μ). Then no cocycle $\beta : \Gamma \times X \to G$ can be simultaneously equivalent to both a diffuse cocycle and a narrow cocycle.*

Proof. Otherwise we may find diffuse

$$\beta_0 : \Gamma \times X \to G$$

and narrow

$$\beta_1 : \Gamma \times X \to G$$

and measurable $\theta : X \to G$ such that a.e.

$$\beta_0(\gamma, x) = \theta(\gamma \cdot x)^{-1} \beta_1(\gamma, x) \theta(x).$$

Claim. For almost every $x \in X$, the function

$$y \mapsto \theta(y)$$

is finite-to-one on $[x]_\Gamma$.

Proof. Otherwise, for all x in a set of positive measure, we may collect together an infinite $\{\gamma_i : i \in \mathbb{N}\} \subseteq \Gamma$ such that there is a single $g_0 \in G$ with

$$\theta(\gamma_i \cdot x) \equiv g_0.$$

We may apply narrowness of β_1 to assume that $\beta_1(\gamma_i, x) \equiv g_1$ some single g_1, and obtain that for every i,

$$\beta_0(\gamma_i, x) \equiv g_0 g_1 \theta(x),$$

contradicting the diffuseness of β_0.

Now we let $(h_n)_{n\in\mathbb{N}}$ enumerate Γ and at each x we let $n(x)$ be least such that there exists $y \in [x]_\Gamma$ with $\theta(y) = h_{n(x)}$. Note that this is measurable and Γ-invariant, and so if we assign to each x the set $S_x = \{y \in [x]_\Gamma : \theta(y) = h_{n(x)}\}$ then we obtain a measurable and invariant selection of a finite subset of each equivalence class. Taking a Borel linear ordering on X and selecting for each x the least element of S_x we obtain a selector with the usual contradiction to the invariance of the measure. ⊣

For the next lemma we need to know that $\mathbb{Z}_p * \mathbb{Z}$ is near hyperbolic. This can be seen either by a direct proof along the lines of C3 from the appendix or an outright appeal to the well known fact that this group is actually hyperbolic and hence near hyperbolic by C4.4.

Lemma 3.11. *Suppose that $\Gamma = F_2 \times F_2$ acts freely and in a Borel way on a standard Borel space X with an invariant probability measure μ, so that the action of each non-amenable subgroup is E_0-ergodic and the action of each infinite subgroup is ergodic. Let $\bar{H}_2 = \mathbb{Z}_p * \mathbb{Z}$ and let $\bar{H} = F_2 \times \bar{H}_2$. Let*

$$\alpha : \Gamma \times X \to \bar{H}$$

be a Borel cocycle such that its restriction to any non-amenable subgroup is not equivalent to a cocycle taking values in an amenable subgroup of $\bar{H}$. Then there is an infinite cyclic subgroup Δ of Γ such that the restricted cocycle

$$\alpha : \Delta \times X \to \bar{H} = F_2 \times \bar{H}_2$$

is equivalent to a cocycle

$$\hat{\alpha} : \Delta \times X \to \bar{H} = F_2 \times \bar{H}_2,$$

such that

(i) $p_1 \circ \hat{\alpha} : \Delta \times X \to F_2$ is given by a non-trivial homomorphism into F_2;
(ii) $p_2 \circ \hat{\alpha} : \Delta \times X \to \bar{H}_2$ takes its image inside $\mathbb{Z}_p$ a.e.

Proof. We let $F_2 = \langle a, b\rangle$ and consider the induced cocycles

$$\alpha_{11} : (F_2 \times \langle a\rangle) \times X \to F_2,$$

$$\alpha_{21} : (\langle a\rangle \times F_2) \times X \to F_2,$$

obtained by restricting α to the indicated subgroups and composing with the projection of $\bar{H}$ onto its first coordinate. The proof splits here into cases.

Case(1). One of α_{11}, α_{21} is equivalent to a homomorphism whose image in F_2 is non-abelian.

We may then assume without damage to the generality of our argument that there is a homomorphism $\pi : F_2 \times \langle a \rangle \to F_2$ with non-abelian image such that almost everywhere

$$\alpha_{11}(\gamma, x) = \pi(\gamma).$$

Note then that restriction of α_{11}, and hence also α_{21}, to $(\langle a \rangle \times \{1\}) \times X$ is diffuse.

We now apply the dichotomy of Theorem 2.2 to the cocycle

$$\alpha_{21} : (\langle a \rangle \times F_2) \times X \to F_2$$

and the normal cyclic subgroup $\Delta = \langle a \rangle \times \{1\}$ of $\langle a \rangle \times F_2$. By 3.10 we must have that alternative (ii) of 2.2 is excluded, and thus we have (i).

At this stage we consider the cocycle

$$\alpha_{22} : (\langle a \rangle \times F_2) \times X \to \bar{H}_2,$$

obtained by composing the restriction of α with the projection onto the second coordinate of H. We again apply 2.2 to this cocycle and the amenable subgroup $\Delta = \langle a \rangle \times \{1\}$. Alternative (i) of 2.2 would give that the restricted cocycle

$$\alpha_2 : (\langle a \rangle \times F_2) \times X \to \bar{H}$$

is equivalent to a cocycle into an amenable subgroup.

Therefore, we must be in case (ii) of 2.2. We have then a cocycle $\beta : (\langle a \rangle \times F_2) \times X \to \bar{H}_2$ equivalent to α_2 and a finite subgroup F_0 of $H_2 = \mathbb{Z}_p * \mathbb{Z}$ with $\beta(x, \delta) \in F_0$, for all $\delta \in \Delta$. It is an easy combinatorial fact (proved for instance by applying Kurosh's theorem as in the proof of 3.1) that F_0 must be cyclic with generator of the form

$$gug^{-1}$$

for some $u \in \mathbb{Z}_p, g \in H_2$. Replacing β by the equivalent cocycle $(\gamma, x) \mapsto g^{-1}\beta(\gamma, x)g$ we obtain the conclusion of the lemma for the cocycle

$$\hat{\alpha} : \Delta \times X \to F_2 \times \bar{H}_2$$

$$((\langle a^\ell \rangle, 1), x) \to (\pi(a^\ell, 1), g^{-1}\beta(a^\ell, x)g).$$

Case(2). Neither of α_{11}, α_{21} is equivalent to a homomorphism whose image in F_2 is non-abelian. And hence, by 2.2, both are equivalent to cocycles into amenable subgroups.

We will complete the proof of the lemma by gradually arguing that this case splits into a sequence of subcases, each of which in turn renders a contradiction.

We first consider the induced cocycle

$$\alpha_{12} : (F_2 \times \langle a \rangle) \times X \to \bar{H}_2.$$

Applying 2.2, and the lemma hypothesis as in the proof for Case(1), we can find an equivalent cocycle

$$\beta_0 : (F_2 \times \langle a \rangle) \times X \to \bar{H}_2$$

such that

$$x \mapsto \mathbb{Z}_p \beta_0((\gamma, 1), x)\mathbb{Z}_p$$

is constant a.e. for any $\gamma \in F_2$. Then for each γ in F_2 we may choose $h_\gamma \in \bar{H}_2$ such that

(a) h_γ has the form $h_\gamma = n_1\bar{\ell}_1 n_2 ... n_k\bar{\ell}_k n_{k+1}$, some k, each $n_i \in \mathbb{Z}$, each $\bar{\ell}_i \in \mathbb{Z}_p$, and

(b) $\mathbb{Z}_p \beta_0((\gamma, 1), x)\mathbb{Z}_p = \mathbb{Z}_p h_\gamma \mathbb{Z}_p$ a.e.

In other words, we strip away from the ends of $\beta_0(\gamma, x)$ any elements of $\mathbb{Z}_p$.

Subcase(2a). There is some h_γ, for $\gamma \in F_2$, which is not torsion.

Thus the cocycle obtained by restricting β_0 to $(\langle\gamma\rangle \times \{1\}) \times X$ is diffuse. We then go across, as in the proof of Case(1), and consider the cocycle on the other side,

$$\alpha_{22} : (\langle\gamma\rangle \times F_2) \times X \to \bar{H}_2.$$

Again, following 3.10, we have that α_{22} restricted to $(\langle\gamma\rangle \times \{1\})$ is not equivalent to a narrow cocycle. We again apply 2.2, to conclude that α_{22} must be equivalent to a cocycle into an amenable subgroup of H_2, and hence the restricted cocycle

$$(\{1\} \times F_2) \times X \to \bar{H}$$

$$((1, \sigma), x) \mapsto (\alpha_{21}((1, \sigma), x), \alpha_{22}((1, \sigma), x)$$

is equivalent to a cocycle into an amenable subgroup, with a contradiction to the assumptions of the lemma.

Subcase(2b). Whenever $\gamma \in F_2$, h_γ is torsion.

And we split again.

Subsubcase(2bi). $\{h_\gamma : \gamma \in F_2\}$ is finite.

Then, to each x, we may assign the finite set $S_x = \{\beta_0((\gamma, 1), x) : \gamma \in F_2\}$ and observe that $xE_{F_2\times\{1\}}y$ always implies that S_x is an $\bar{H}_2$-translation of S_y, and in fact if $y = (\gamma, 1) \cdot x$ then

$$S_y = \beta_0((\gamma, 1), x)S_x.$$

Since the action of $\bar{H}_2$ on its finite sets has only countably many orbits and the action of $F_2 \times \{1\}$ is ergodic, we can apply the usual cocycle reduction and obtain a single, finite, non-empty $S \subseteq \bar{H}_2$ and a cocycle β_1 equivalent to β_0, with

$$\beta_1((\gamma, 1), x) \cdot S = S.$$

all $\gamma \in F_2$, a.e. $x \in X$. Since the stabilizer of S is a finite subgroup of $\bar{H}_2$, and therefore in particular amenable, we obtain that α restricted to $F_2 \times \{1\}$ is equivalent to

$$((\gamma, 1), x) \mapsto (\alpha_{11}((\gamma, 1), x), \beta_1((\gamma, 1), x)),$$

which in turn, by the case assumptions, is equivalent to a cocycle into an amenable subgroup of H, with a contradiction to assumptions of lemma.

Subsubcase(2bii). $\{h_\gamma : \gamma \in F_2\}$ is infinite.

Recalling our subcase assumptions we have that each h_γ is torsion, and hence by considering the combinatorics of $\mathbb{Z}_p * \mathbb{Z}$ (or applying Kurosh's theorem as at C3.1), we have

$$h_\gamma = g_\gamma u_\gamma g_\gamma^{-1},$$

where $u \in \mathbb{Z}_p$ and $g_\gamma = n_{\gamma,1}\bar{\ell}_{\gamma,1}n_{\gamma,2}...\bar{\ell}_{\gamma,k(\gamma)-1}n_{\gamma,k(\gamma)}$, where $k(\gamma) \in \mathbb{N}$, each $n_{\gamma,i} \in \mathbb{Z}$ non-zero, assuming $g_\gamma \neq 1$, and each $\bar{\ell}_{\gamma,i} \in \mathbb{Z}_p$ non-zero. We will forthwith remember this notation, and say that $k(\gamma)$ is *the length of* h_γ.

Claim. If $g_{\gamma_1}, g_{\gamma_2} \neq 1$ and the length of h_{γ_1} is no greater than the length of h_{γ_2}, then for every $i \leq \gamma_1(k)$

$$n_{\gamma_1,i} = n_{\gamma_2,i},$$

and for $i < \gamma_1(k)$

$$\bar{\ell}_{\gamma_1,i} = \bar{\ell}_{\gamma_2,i}.$$

Proof. Otherwise, a moment's consideration of the combinatorics of concatenating words in $\mathbb{Z}_p * \mathbb{Z}$, shows that for any $w_1, w_2, w_3 \in \mathbb{Z}_p$,

$$w_1 h_{\gamma_1} w_2 h_{\gamma_2} w_3$$

is of infinite order. But then since each $\beta_0((\gamma_1\gamma_2, 1), x)$ has this form for a.e. x, we obtain a contradiction to the subcase assumptions.

Appealing to the subsubcase assumptions, we have that a.e.

$$\{\beta_0((\gamma, 1), x) : \gamma \in F_2\} \subset \{w_1 h_\gamma w_2 : \gamma \in F_2, w_1, w_2 \in \mathbb{Z}_p\}$$

is infinite. The last claim implies that its closure under subsequences in the Cayley graph of H_2, which we henceforth denote by T_x, has at most $2p$ incompatible elements; this in turn implies that ∂T_x has a finite, non-zero, number of infinite branches.

Therefore, in the notation of appendix C3.1, we have an assignment

$$x \mapsto E_x,$$
$$X \to [S_{\bar{H}_2}]^{<\infty},$$

of finite subsets of the reduced words from $\mathbb{Z}_p \sqcup \mathbb{Z}$, such that for all $\gamma \in F_2$, a.e. $x \in X$

$$E_{(\gamma,1)\cdot x} = \beta_0((\gamma, 1), x) \cdot E_x.$$

Since the shift action of $\bar{H}_2$ on $[S_{\bar{H}_2}]^{<\infty}$ gives rise to an equivalence relation which is hyperfinite (since it can be viewed as the tail equivalence relation, just as in the proof of C3.1 i), and the action of $F_2 \times \{1\}$ on X is E_0-ergodic, we may apply the usual cocycle reduction to find a cocycle β_1 which is equivalent to β_0 and a single $E \in [S_{\bar{H}_2}]^{<\infty}$ such that a.e.

$$\beta_1((\gamma, 1), x) \cdot E = E.$$

Then the restriction of α to $(F_2 \times \{1\}) \times X$ is equivalent to the cocycle

$$(F_2 \times \{1\}) \times X \to \bar{H}$$
$$((\gamma, 1), x) \mapsto (\alpha_{21}((\gamma, 1), x), \beta_1((\gamma, 1), x),$$

thus equivalent to a cocycle into an amenable subgroup, with a contradiction to the assumptions of the lemma. ⊣

Corollary 3.12. *Suppose that $\Gamma = F_2 \times F_2$ acts freely and in a Borel way on a standard Borel space X with an invariant measure μ, so that the action of each non-amenable subgroup is E_0-ergodic and the action of each infinite subgroup is ergodic. Let $\bar{H}_2 = \mathbb{Z}_p * \mathbb{Z}$ and let $\bar{H} = F_2 \times \bar{H}_2$ act freely and in a Borel way on a standard Borel space Y. Suppose $\rho : X \to Y$ witnesses $E_\Gamma^X \leq_B E_{\bar{H}}^Y$.*

Then there is an infinite cyclic subgroup Δ_0 of $\bar{H}$ and a Δ_0-invariant measure ν on Y with

$$\nu(\rho(X)) \neq 0.$$

Proof. Let α be the cocycle associated to the reduction ρ. Applying the last lemma we may assume that there is an infinite cyclic $\Delta \leq \Gamma$ such that:

(i) $p_1 \circ \alpha : \Delta \times X \to F_2$ is given by a non-trivial homomorphism $\pi : \Delta \to F_2$;

(ii) $p_2 \circ \alpha : \Delta \times X \to \bar{H}_2$ has image inside $\mathbb{Z}_p$ a.e.

Then let $\Delta_0 = \pi(\Delta)$, and note that the actions of Δ_0 and $\mathbb{Z}_p$ on Y commute. Since $\mathbb{Z}_p$ is finite, we may assume that $Y = Y_0 \times \mathbb{Z}_p$ and that $\mathbb{Z}_p$ acts by permuting the second coordinate of Y. We first obtain a suitably invariant measure on Y_0. Defining the function

$$\varphi : X \to Y_0$$

by

$$\varphi(x) = y$$

if and only if $\rho(x) = (y, \bar{\ell})$ for some $\bar{\ell}$, we can define the measure $\nu_0 = \varphi_* \mu$ by

$$\nu_0(A) = \mu(\varphi^{-1}(A)).$$

Since φ respects the homomorphism $\pi : \Delta \to \Delta_0$, we obtain a Δ_0 invariant measure on Y by

$$\nu(A) = \int_{Y_0} |\{\bar{\ell} : (y, \bar{\ell}) \in A\}| d\nu_0(y).$$

$\dashv$

On the basis of these facts, we can complete the proof as before. The argument in Lemma 3.6 goes through untouched for $\bar{H}_2$, since this group cannot be mapped homomorphically onto a non-abelian free group. The rest of this alternate proof of 3.8 goes then as follows:

Let $\Gamma = F_2 \times F_2$ and $\bar{H} = F_2 \times \bar{H}_2$. Since $\bar{H}_2$ contains a copy of F_2, we view Γ as a subgroup of $\bar{H}$. We consider the free part $(2)^{\bar{H}}$ of the shift action of $\bar{H}$ on $2^{\bar{H}}$ and the usual product measure. We recall from A6.1 that this action is mixing, so every infinite subgroup acts ergodically, and from A4.1 that every non-amenable subgroup acts E_0-ergodically. Fix an $\bar{H}$-invariant Borel subset X of $(2)^{\bar{H}}$ such that $\mu(X) = 1$ and every infinite cyclic subgroup of H acts uniquely ergodically on X. We also write μ for the restriction of μ to X.

We now let

$$E = E_\Gamma^X, F = E_{\bar{H}}^X.$$

Clearly $E \subseteq F$. So it is enough to check that $E \not\leq_B F$.

Suppose instead $\rho : X \to X$ is a Borel reduction of E to F. Applying 3.12, we obtain that there is an infinite cyclic subgroup Δ_0 of H and a Δ_0-invariant measure ν on Y with $\nu(\rho(X)) \neq 0$. Appealing to unique ergodicity, we have $\nu = \mu$, and hence

$$\mu(\rho(X)) \neq 0,$$

so

$$\mu(\Delta_0 \cdot (\rho(X))) = 1.$$

We let

$$Y = \bigcap_{h \in \bar{H}} h \cdot \Delta_0 \cdot (\rho(X)),$$

and note that Y is H-invariant and still of μ measure 1; we let μ denote the restriction of μ to Y.

For each $y \in Y$, we may in a Borel manner assign some $\theta(y) \in X$ with $\rho(\theta(y)) E_{\bar{H}}^X y$. But then θ witnesses

$$E_{\bar{H}}^Y \leq_B E_\Gamma^X,$$

which contradicts 3.6. $\dashv$

3D. Relative ergodicity and rigidity results for product group actions

Theorem 3.13. *Let $\Gamma = \Gamma_1 \times \cdots \times \Gamma_n \times \Delta, n \geq 1$, be a countable group, where Δ is an infinite amenable group, and let $H = H_1 \times \cdots \times H_n$, where each H_i is countable and amenable or free. Let Γ act in a Borel way on a standard Borel space X with invariant measure μ and assume that Δ acts ergodically and each Γ_i acts E_0-ergodically. Let $\alpha : \Gamma \times X \to H$ be a Borel cocycle. Then one of the following holds:*

(i) There is $\emptyset \neq \{i_1, \ldots, i_k\} \subseteq \{1, \ldots, n\}$ such that $\alpha|((\Gamma_{i_1} \times \cdots \times \Gamma_{i_k} \times \Delta) \times X) \sim \beta$, with $\beta((\Gamma_{i_1} \times \cdots \times \Gamma_{i_k} \times \Delta) \times X) \subseteq H_0 \leq H$, where H_0 is amenable.

(ii) For some $1 \leq i \leq n, 1 \leq j \leq n$, we have that $\alpha|((\Gamma_i \times \Delta) \times X) \sim \beta$, with $\beta((\Gamma_i \times \Delta) \times X) \subseteq H_j$.

Corollary 3.14 *Let Γ, H, X be as in the preceding theorem. Then if H acts freely in a Borel way on Y, E_Γ^X is E_H^Y-ergodic.*

In particular, the shift action of $\Gamma = \Gamma_1 \times \cdots \times \Gamma_n \times \mathbb{Z}$ on 2^Γ, where Γ_i are not amenable, is E-ergodic, for any equivalence relation E induced by a free Borel action of the product of n amenable or free groups, e.g., F_2^n or $F_2^{n-1} \times \mathbb{Z}$, etc.

Proof of 3.14. Suppose H acts freely in a Borel way on Y and $\rho : X \to Y$ is a Borel homomorphism of E_Γ^X to E_H^Y. Let α be the associated cocycle,

$$\alpha(\gamma, x) \cdot \rho(x) = \rho(\gamma \cdot x), \ \gamma \in \Gamma.$$

If (i) of Theorem 3.13 holds, we can let $x \mapsto h_x$ be a Borel map from X to H, so that for $\gamma \in \Gamma_{i_1} \times \cdots \times \Gamma_{i_k} \times \Delta$,

$$\alpha(\gamma, x) = h_{\gamma \cdot x} \beta(\gamma, x) h_x^{-1}, \ \mu\text{-a.e. } (x).$$

Then if

$$\sigma(x) = h_x^{-1} \cdot \rho(x),$$

σ is also a Borel homomorphism of E_Γ^X to E_H^Y and its associated cocycle restricted to $\Gamma_{i_1} \times \cdots \times \Gamma_{i_k} \times \Delta$ is equal to $\beta(\gamma, x)$, μ-a.e. (x), so it takes values as an amenable group $H_0 \leq H$, μ-a.e. It follows that σ, restricted to a co-null set, is a homomorphism of $E_{\Gamma_{i_1}}^X$ to a hyperfinite subequivalence relation of E_H^Y (since an equivalence relation induced by a Borel action of an amenable group is hyperfinite a.e. for any measure, by [OW]). Since Γ_{i_1} acts E_0-ergodically, σ maps into a single E_H^Y-class, thus so does ρ.

So let us assume that (ii) of Theorem 3.13 holds. Then let $x \mapsto f_x$ be Borel from X to H, so that for $\gamma \in \Gamma_i \times \Delta$,

$$\alpha(\gamma, x) = f_{\gamma \cdot x} \beta(\gamma, x) f_x^{-1}, \ \mu\text{-a.e. } (x).$$

Put as usual

$$\tau(x) = f_x^{-1} \cdot \rho(x),$$

so that τ is a Borel homomorphism of E_Γ^X to E_H^Y, whose associated cocycle α' satisfies, for $\gamma \in \Gamma_i \times \Delta$,

$$\alpha'(\gamma, x) = \beta(\gamma, x) \ \mu\text{-a.e. } (x),$$

so $\alpha'(\gamma, x)$ takes values in H_j μ-a.e. (x). If H_j is amenable, then we can see that ρ maps into a single E_H^Y-class as in the previous case. So let us assume that H_j is free. Then we can apply Theorem 2.2 to the action of $\Gamma_i \times \Delta$ on X and the cocycle $\alpha'|((\Gamma_i \times \Delta) \times X)$, which takes values in H_j μ-a.e. If case (i) of 2.2 applies, we are done, as before, so we can assume that we are in case (ii)$'$ of 2.2. Then $\alpha'|(\Gamma_i \times \Delta) \times X$ is equivalent to a homomorphism π of $\Gamma_i \times \Delta$ into H_j, which is trivial on Δ. As usual, we can change τ to a Borel homomorphism φ of $E^X_{\Gamma_i \times \Delta}$ into E^Y_H such that the cocycle of φ is equal to π, μ-a.e., and $\varphi(x) E^Y_H \tau(x)$, μ-a.e. (x). But then φ is Δ-invariant, so by the ergodicity of the Δ-action, it is constant a.e., and thus τ and so ρ maps into a single E^Y_H-class, a.e. ⊣

Proof of Theorem 3.13.

Write $\alpha = (\alpha_1, \dots, \alpha_n)$, where $\alpha_j = p_j \circ \alpha$, with p_j the projection of H onto H_j. Then, applying 2.2 to α_j, we see that we have one of two possibilities:

(I) $\alpha_j \sim \beta_j$, where β_j takes values into an amenable subgroup of H_j

or

(II) $\alpha_j \sim \pi_j$, where π_j is a homomorphism of Γ into H_j, and π_j is trivial on Δ.

If (I) holds for all $j = 1, \dots, n$, clearly $\alpha \sim \beta$, where β takes values in an amenable subgroup of H, so (i) holds (with $\{i_1, \dots, i_k\} = \{1, \dots, n\}$).

Otherwise (II) holds for some j, with π_j not taking values into an amenable subgroup of H_j. If $n = 1$, clearly (ii) of 3.13 holds, so we assume $n > 1$. Also obviously H_j is free. We now use the following standard fact about free groups (see, e.g., [LS, 2.18]): The relation $a \square b \Leftrightarrow ab = ba$, in any free group F, is an equivalence relation on the set of non-trivial elements of F. It follows that there is a unique i such that $\pi_j(\Gamma_i)$ is not abelian and $\pi_j(\Gamma_{i'}) = \{1\}$, for $i' \neq i$. Then $\alpha|((\Gamma_1 \times \cdots \times \hat{\Gamma}_i \times \cdots \times \Gamma_n \times \Delta) \times X) \sim \beta$, with β taking values in $H_1 \times \cdots \times \hat{H}_j \times \cdots \times H_n$, where $\hat{\Gamma}_i, \hat{H}_j$ means that the corresponding groups are omitted. Proceeding this way by induction, we either get $\emptyset \neq \{i_1, \dots i_k\} \subseteq \{1, \dots, n\}$, so that $\alpha|(\Gamma_{i_1} \times \cdots \times \Gamma_{i_k} \times \Delta) \times X$ is equivalent to a cocycle taking values in an amenable subgroup of H or else this procedure continues for $n - 1$ steps and we find some i, j so that $\alpha|(\Gamma_i \times \Delta) \times X$ is equivalent to a cocycle taking values in H_j, so we are done. ⊣

We also have the following rigidity result concerning the shift action of product groups. (We will see stronger results in §5.)

Theorem 3.15. *Let Δ_0 be a countable, infinite amenable group, H_0 a countable non-amenable group, Δ_1 a countable amenable group and H_1 a torsion-free near-hyperbolic countable group. Then if $F(H_0 \times \Delta_0, 2) \leq_B F(H_1 \times \Delta_1, 2)$, there is a normal amenable subgroup $N \trianglelefteq H_0$ such that H_0/N is (isomorphic to) a subgroup of H_1. In particular, if H_0 is torsion-free, hyperbolic,*

$$F(H_0 \times \Delta_0, 2) \leq_B F(H_1 \times \Delta_1, 2)$$

iff H_0 is (isomorphic to) a subgroup of H_1.

Proof. Denote by μ the usual product measure on $2^{H_0 \times \Delta_0}$ which concentrates on $(2)^{H_0 \times \Delta_0}$. Suppose $\rho : (2)^{H_0 \times \Delta_0} \to (2)^{H_1 \times \Delta_1}$ is a Borel reduction of $F(H_0 \times \Delta_0, 2)$ to $F(H_1 \times \Delta_1, 2)$. Let $\alpha : H_0 \times \Delta_0 \to H_1 \times \Delta_1$ be the associated cocycle. Let α_1 be its projection to H_1. Then by 2.2, for $H_0 \times \Delta_0$ and α_1, we have two possibilities:

(i) α_1 is equivalent to a cocycle taking values in an amenable subgroup of H_1 and thus α is equivalent to a cocycle β taking values in an amenable subgroup G

of $H_1 \times \Delta_1$. Let then σ be a reduction from $F(H_0 \times \Delta_0, 2)$ to $F(H_1 \times \Delta_1, 2)$, whose associated cocycle is β, μ-a.e. Then, μ-a.e., σ reduces $F(H_0 \times \Delta_0, 2)$ to ${E_G^{(2)}}^{H_1 \times \Delta_1}$ and ${E_G^{(2)}}^{H_1 \times \Delta_1}$ is hyperfinite, ν-a.e., where $\nu = \sigma_* \mu$. Thus $F(F_0 \times \Delta_0, 2)$ is hyperfinite, μ-a.e., so $H_1 \times \Delta_1$ is amenable, a contradiction.

(ii) There is a homomorphism $\pi : H_0 \times \Delta_0 \to H_1$, which is trivial on Δ_0 and $\alpha_1 \sim \pi$. It is thus enough to verify that $N = \ker(\pi)$ is amenable. If $\alpha = (\alpha_1, \alpha_2)$, where α_2 is the projection to Δ_1, then $\alpha \sim (\pi, \alpha_2)$, and we can find a reduction τ of $F(H_0 \times \Delta_0, 2)$ to $F(H_1 \times \Delta_1, 2)$, whose associated cocycle is (π, α_2), μ-a.e. Then note that τ is also a reduction of ${E_N^{(2)}}^{H_0 \times \Delta_0}$ to ${E_{\Delta_1}^{(2)}}^{H_1 \times \Delta_1}$, which is hyperfinite λ-a.e., where $\lambda = \tau_* \mu$. It follows that ${E_N^{(2)}}^{H_0 \times \Delta_0}$ is hyperfinite, μ-a.e., so N is amenable.

The final conclusion follows from C4.2. ⊣

CHAPTER 4

Factoring Homomorphisms

We will start with a refinement of Theorem 2.2, which deals with the case when the action of Γ is not necessarily E_0-ergodic. In stating this result it is convenient to introduce the following:

Definition 4.1. *A near-hyperbolic group H is called* nice *if in the definition 2.1, (a) is strengthened by requiring that the action of H on the set of elements of $\mathcal{M}_{\leq 2}(K)$ with non-trivial stabilizers has corresponding equivalence relation tame.*

By Appendices C2.2, C4.3 this is true for all free groups and more generally all torsion-free hyperbolic groups. $\mathbb{Z}_2 * \mathbb{Z}_3$ is, however, an example of a non-nice, near-hyperbolic group.

Theorem 4.2. *Let Γ be a countable group and $\Delta \trianglelefteq \Gamma$ an infinite normal amenable subgroup. Suppose Γ acts in a Borel way on the standard Borel space X with invariant measure μ. Assume that the action of Δ is ergodic. Let H be a nice near-hyperbolic group and let $\alpha : \Gamma \times X \to H$ be a Borel cocycle. Then one of the following three possibilities holds:*

(i) There is a Borel cocycle $\beta : \Gamma \times X \to H$ equivalent to $\alpha, \alpha \sim \beta$, such that $\beta(\Gamma \times X) \subseteq H_0$, where $H_0 \leq H$ is an amenable subgroup of H.

(i) There is a conull Borel Γ-invariant $X_0 \subseteq X$, a free Borel action of H on a standard Borel space Y_0 with $E_H^{Y_0}$ hyperfinite, and a Borel homomorphism $\rho : E_\Gamma^{X_0} \to E_H^{Y_0}$ such that for all $\gamma \in \Gamma, x \in X_0$*

$$\alpha(\gamma, x) = 1 \Leftrightarrow \rho(x) = \rho(\gamma \cdot x).$$

(ii) There is a Borel cocycle $\beta : \Gamma \times X \to H$ equivalent to $\alpha, \alpha \sim \beta$, such that $\beta(\Delta \times X) \subseteq F_0$, where $F_0 \leq H$ is a finite subgroup of H. And in this case, the double coset

$$\pi(\gamma) = F_0 \beta(\gamma, x) F_0, \ \gamma \in \Gamma,$$

depends only on γ, μ-a.e. (x).

In particular, if H is torsion-free, then (ii) can be replaced by:

(ii)′ There is a homomorphism $\pi : \Gamma \to H$ with $\pi(\Delta) = \{1\}$ such that π (viewed on cocycle from $\Gamma \times X$ into H, via $\pi(\gamma, x) = \pi(\gamma)$) is equivalent to $\alpha, \alpha \sim \pi$.

Proof. We use the notation and follow the argument in the proof of 2.2, except that we also assume that if

$$Z = \{\nu \in \mathcal{M}_{\leq 2}(K) : \nu \text{ has non-trivial stabilizer}\},$$

then E_H^Z is tame. Let $Y = \mathcal{M}_{\leq 2}(K) \setminus Z$.

Case (ii) is identical to that of 2.2, so we analyze further case (i) of 2.2, in order to obtain (i) or (i)*.

Because Γ acts ergodically on X, we have that

$$\rho(x) \in Z, \ \mu\text{-a.e. } (x)$$

or

$$\rho(x) \in Y, \ \mu\text{-a.e. } (x)$$

In the first case, we obtain (i), exactly as in the proof of 2.2, noting that since E_H^Z is tame we only need the ergodicity of the Γ-action. In the second case, let $\lambda = \rho_* \mu$ and find a Γ-invariant Borel set $X_0 \subseteq X$ and an H-invariant Borel set $Y_0 \subseteq Y$ with $\mu(X_0) = 1, \rho(X_0) = Y_0 \subseteq Y$, $E_H^{Y_0}$ hyperfinite, and $\alpha(\gamma, x) \cdot \rho(x) = \rho(\gamma \cdot x)$, for $\gamma \in \Gamma, x \in X_0$. Then $\rho : E_\Gamma^{X_0} \to E_H^{Y_0}$ is a Borel homomorphism, H acts freely on Y_0, and for $\gamma \in \Gamma, x \in X_0$

$$\alpha(\gamma, x) = 1 \Leftrightarrow \rho(x) = \rho(\gamma \cdot x),$$

so the proof is complete. ⊣

We next define the notion of factoring for Borel homomorphisms.

Definition 4.3. *Let E, F be countable Borel equivalence relations on X, Y, resp. Let $\rho : X \to Y$ be a Borel homomorphism of E to F. We say that ρ* factors through *a countable Borel equivalence relation R on Z if there are Borel homomorphisms $\sigma : X \to Z, \tau : Z \to Y$ of E to R, and R to F, resp., with $\rho = \tau \circ \sigma$.*

We now have the following result which we will use in conjunction with Theorem 4.2 to obtain rigidity theorems for actions of product groups.

Theorem 4.4. *Let Γ be a countable group acting in a Borel way on a standard Borel space X. Let $H_1, \dots, H_n$ be countable groups and assume that $H = H_1 \times \cdots \times H_n$ acts freely in a Borel way on a standard Borel space Y. Let $\rho : X \to Y$ be a Borel homomorphism of E_Γ^X to E_H^Y, with associated cocycle α, i.e., $\alpha(\gamma, x) \cdot \rho(x) = \rho(\gamma \cdot x)$. Let p_i be the projection of H to H_i and put $\alpha_i = p_i \circ \alpha, i = 1, \dots, n$. Suppose that for each i, one of the following holds:*

(i) $\alpha_i(\Gamma \times X) \subseteq H_i'$, where H_i' is an amenable subgroup of H_i.

(ii) There is a hyperfinite Borel equivalence relation E_i on Z_i and a Borel homomorphism $\rho_i : X \to Z_i$ of E_Γ^X to E_i such that for $\gamma \in \Gamma, x \in X$:

$$\alpha_i(\gamma, x) = 1 \Leftrightarrow \rho_i(\gamma \cdot x) = \rho_i(x).$$

Then there is a countable Borel equivalence relation $\hat{F}$ on $\hat{Z}$ such that ρ factors through $\hat{F}$, and $\hat{F}$ is $\hat{\mu}$-hyperfinite, for any measure $\hat{\mu}$ on $\hat{Z}$.

Proof. We will prove the result here in the case ρ is a Borel reduction, which is enough for the applications in §5. We will give the proof in the general case, where ρ is an arbitrary Borel homomorphism, in Appendix E, see E2.1.

We will also assume, for convenience, that the Continuum Hypothesis (CH) is true. This is harmless, since the proposition we want to prove is a projective statement, so CH can be avoided by standard metamathematical results (see, e.g., Appendix 2 in [AL], where this is explained in some detail). Since ρ is a reduction, clearly ρ is countable-to-1, so $\rho[X]$ is Borel and thus the E_H^Y-saturation of $\rho(X)$ is Borel. So we can assume without loss of generality that $\rho(X)$ is a complete section for E_H^Y, i.e., it meets every E_H^Y-class.

It is clear that it is enough to prove the result for $n = 2$. Then there are three different cases depending on whether (i) holds for both α_1, α_2, (ii) holds for both α_1, α_2, or (i) holds for one α_i and (ii) for the other.

In the first case, let $H' = H_1' \times H_2' \leq H$, so that H' is amenable. Let $\hat{Z} = Y, \hat{F} = E^Y_{H'}$. Then clearly $\hat{F}$ is μ-hyperfinite for all μ (by [OW]), as it is induced by a Borel action of an amenable group, and of course ρ factors through $\hat{F}$.

The proofs in the last two cases are quite similar, so we will give below the argument for the third case. So let us assume that $\alpha_1(\Gamma \times X) \subseteq H_1' \leq H_1$, with H_1' amenable, and that there is a hyperfinite Borel equivalence relation E_2 on Z_2 and a Borel homomorphism $\rho_2 : X \to Z_2$ of E^X_Γ to E_2 such that for $\gamma \in \Gamma, x \in X$:

$$\alpha_2(\gamma, x) = 1 \Leftrightarrow \rho_2(\gamma \cdot x) = \rho_2(x).$$

Since clearly ρ is a homomorphism of E^X_Γ to $E^Y_{H_1' \times H_2}$, we may as well assume that H_1 itself is amenable. We will then show that we can take $\hat{Z} = Y, \hat{F} = E^Y_H$. Since clearly ρ factors through $\hat{F}$, it is enough to show that E^Y_H is μ-hyperfinite for each μ. By [CFW], and the discussion in [JKL, Sections 2.2, 2.3], it is enough to show that E^Y_H is measure-amenable, which means that there is a universally measurable (in the sense of [JKL, Def. 2.7]) map $C \mapsto \varphi_C$ which assigns to each E^Y_H-class C a mean φ_C on C.

Let $F_1 = E^Y_{H_1}, F_2 = E^Y_{H_2}$, and, since E_2 is hyperfinite, fix a Borel relation $\prec$ on Z_2 which induces on each E_2-class an ordering isomorphic to $\mathbb{Z}$ or else finite.

Now fix $y, z \in \rho(X)$ with yF_1z and take any $x_1, x_2 \in X$ with $\rho(x_1) = y, \rho(x_2) = z$. Then there is $\gamma \in \Gamma$, with $\gamma \cdot x_1 = x_2$ and, since $\alpha_2(\gamma, x_1) = 1$, it follows that $\rho_2(x_1) = \rho_2(x_2)$. Put, for any $y \in \rho(X)$,

$$\sigma(y) = \rho_2(x), \text{ for any } x \in X \text{ with } \rho(x) = y.$$

The above shows that σ is well-defined, Borel (since it has analytic graph), and F_1-invariant. Also if $y, z \in \rho(X), yE^Y_Hz$, and it is not the case that yF_1z, then if $x_1, x_2 \in X, \gamma \in \Gamma$, are such that $\rho(x_1) = y, \rho(x_2) = z$ and $\gamma \cdot x_1 = x_2$, then $\alpha_2(\gamma, x_1) \neq 1$, so $\rho_2(x_1) \neq \rho_2(x_2)$, thus $\sigma(y) \neq \sigma(z)$. For any E^Y_H-class C, define then the following ordering $\prec_C$ on the set of classes $[y]_{F_1}$, with $y \in \rho(X) \cap C$:

$$[y]_{F_1} \prec_C [z]_{F_1} \Leftrightarrow \sigma(y) \prec \sigma(z).$$

By the above, this is well defined, and $\prec_C$ has order type $\mathbb{Z}, \mathbb{N}, \mathbb{N}^*$ (= the reverse ordering in $\mathbb{N}$) or finite. We will assume that $\prec_C$ has always order type $\mathbb{Z}$, since the other cases can be handled in a similar, but easier way.

Since H_1 is amenable, and $F_1 = E^Y_{H_1}, F_1$ is measure-amenable (see, e.g., the paragraph following [JKL, Theorem 2.8]), so fix a universally measurable map $C_1 \mapsto \psi_{C_1}$, where ψ_{C_1} is a mean on C_1, and C_1 varies over F_1-classes. We finally define a universally measurable map $C \mapsto \varphi_C$, where C varies over E^Y_H-classes, with φ_C a mean on C.

Let θ be a universally measurable invariant mean on $\mathbb{Z}$ (see, e.g., [JKL, Cor. 2.2]). Then for each E^Y_H-class C, we define φ_C, a mean on C, as follows:

Fix $f = \ell_\infty(C)$. For each F_1-class $C_1 \subseteq C$, let $f_{C_1} = f|C_1 \in \ell_\infty(C_1)$. Let $g(C_1) = \psi_{C_1}(f_{C_1})$. Now fix a bijection $\xi : \mathbb{Z} \to \{[y]_{F_1} : y \in \rho(X) \cap C\}$, so that

$$m < n \Leftrightarrow \xi(m) \prec_C \xi(n).$$

Put, finally

$$\varphi_C(f) = \theta(g \circ \xi).$$

Note here that $g \circ \xi \in \ell_\infty(\mathbb{Z})$ and that $\varphi_C(f)$ is independent of ξ, by the invariance of θ. The verification that $C \mapsto \varphi_C$ is universally measurable is routine and the proof is therefore complete. ⊣

Remark. The preceding proof, for the case when ρ is a reduction, shows without any use of CH or the result of [CFW], that one can take in this case the relation $\hat{F}$ in the statement of 4.4 to be 2-amenable (in the sense of [JKL, Def. 2.12]), and again this is enough for the applications in §5.

We finally note, for further reference, that the argument for the proof of 4.4 also gives the following.

Theorem 4.5 *Let Γ be a countable group acting in a Borel way on a standard Borel space X. Let H be a countable group, $\Delta_1 \trianglelefteq H$ a normal amenable subgroup and put $H_1 = H/\Delta_1$. Assume that H acts in a Borel way on a standard Borel space Y. Let ρ be a Borel reduction of E_Γ^X to E_H^Y with associated cocycle α. Let p_1 be the projection of H to H_1 and put $\alpha_1 = p_1 \circ \alpha$. Suppose that there is a hyperfinite Borel equivalence relation E_1 on Z_1 and a Borel homomorphism $\rho_1 : X \to Z_1$ of E_Γ^X to E_1 such that for $\gamma \in \Gamma, x \in X$:*

$$\alpha_1(\gamma, x) = 1 \Leftrightarrow \rho_1(\gamma \cdot x) = \rho_1(x).$$

Then E_Γ^X is μ-hyperfinite for any measure μ on X.

CHAPTER 5

Further Applications

We will now derive some applications of the results in §4.

5A. Rigidity results for reducibility and stable orbit equivalence

Theorem 5.1. *Let Γ be a countable non-amenable group, $\Delta \trianglelefteq \Gamma$ an infinite amenable normal subgroup and suppose Γ acts freely in a Borel way on a standard Borel space X with invariant measure μ, so that the action of Δ is ergodic. Let H be a countable group, $\Delta_1 \trianglelefteq H$ a normal subgroup such that $H_1 = H/\Delta_1$ is a nice, near-hyperbolic, torsion-free group. Suppose H acts freely in a Borel way on a standard Borel space Y. If $E_\Gamma^X \leq_B E_H^Y$, then there is an amenable normal subgroup $N \trianglelefteq \Gamma$ containing Δ, so that Γ/N is (isomorphic to) a subgroup of H_1.*

Proof. Let $\rho : X \to Y$ be a Borel reduction of E_Γ^X to E_H^Y, let $\alpha : \Gamma \times X \to H$ be the associated cocycle and $\alpha_1 = p_1 \circ \alpha$ its projection to H_1, where $p_1 : H \to H_1$ is the canonical projection. We apply 4.2 to α_1.

If (i) of 4.2 holds, then, by the usual arguments (see Appendix B.1), and neglecting null sets, we see that there is a Borel reduction $\bar{\rho} : X \to Y$ of E_Γ^X to E_H^Y, whose associated cocycle $\bar{\alpha}$ satisfies $\bar{\alpha}(\Gamma \times X) \subseteq G \leq H$, with G amenable. Then $E_\Gamma^X \leq_B E_G^Y$, so E_Γ^X is 1-amenable (in the terminology of [JKL, Section 2.4]), thus by [JKL, 2.15, 2.13, 2.14] Γ is amenable, a contradiction.

If (i)* of 4.2 holds, then, by 4.5, we see that E_Γ^X is μ-hyperfinite, for any measure μ on X, so Γ is again amenable, a contradiction.

Finally, if (ii) of 4.2 holds, or rather (ii)′, since H_1 is torsion-free, we have a homomorphism $\pi : \Gamma \to H_1$ with $N = \ker(\pi) \geq \Delta$ and $\alpha_1 \sim \pi$. Then we can find a reduction $\bar{\rho}$ of E_Γ^X to E_H^Y such that the associated cocycle $\bar{\alpha}$ satisfies $p_1 \circ \bar{\alpha} = \pi$, neglecting null sets. Then $\bar{\alpha}|(N \times X)$ takes values in Δ_1, so $E_N^X \leq_B E_{\Delta_1}^Y$, thus, as before, N is amenable, and π clearly shows that Γ/N is (isomorphic to) a subgroup of H_1. ⊣

Corollary 5.2. *Suppose that a non-amenable countable group H_0 contains no non-trivial normal amenable subgroups (e.g., a non-abelian free group or even any torsion-free hyperbolic group which is not cyclic). Suppose Δ_0 is infinite amenable and $H_0 \times \Delta_0$ acts freely in a Borel way on X with invariant measure μ, such that the Δ_0-action is ergodic. Let H_1 be torsion-free hyperbolic (e.g., a free group), Δ_1 amenable and suppose $H_1 \times \Delta_1$ acts freely in a Borel way on Y. If $E_{H_0 \times \Delta_0}^X \leq_B E_{H_1 \times \Delta_1}^Y$, then H_0 is (isomorphic to) a subgroup of H_1.*

The conclusion of Theorem 5.1 can be strengthened in the case that the actions of Γ on X and H on Y are stably orbit equivalent. More precisely we have:

Theorem 5.3. *In the context of 5.1, assume moreover that the action of H on Y has an invariant measure ν and H' acts ergodically for any non-amenable subgroup $\Delta_1 \leq H' \leq H$. If the action of Γ on X and of H on Y are stably orbit equivalent, then there is an amenable normal subgroup $N \trianglelefteq \Gamma$ containing Δ, so that $\Gamma/N \cong H_1$.*

Proof. In the notation of the proof of 5.1, our extra assumptions imply that we can choose $\rho : X \to Y$, so that for any null set $X_0 \subseteq X, \rho(X_0) \subseteq Y$ is null, and for each co-null set $X_1 \subseteq X$, the E^Y_H-saturation of $\rho(X_1)$ is co-null. By following the proof of 5.1, we see that there is a homomorphism $\pi : \Gamma \to H_1$ with $N = \ker(\pi) \geq \Delta$, and N amenable, and a Borel reduction $\bar{\rho}$ of E^X_Γ to E^Y_H with $\rho(x) E^Y_H \bar{\rho}(x)$, μ-a.e. (x), so that $\bar{\rho}$ also sends null sets to null sets, and co-null sets to sets whose saturation is co-null, and moreover the cocycle $\bar{\alpha}$ associated to $\bar{\rho}$ satisfies (a.e., but we will assume it is everywhere, without causing any harm) $p_1 \circ \bar{\alpha} = \pi$, where $p_1 : H \to H_1$ is the canonical projection. Say $\pi(\Gamma) = H_0 \leq H_1$. We want to show that $H_0 = H_1$. Of course $\bar{\alpha}(\Gamma \times X) \subseteq p_1^{-1}(H_0) = H'$.

Put $\bar{\rho}(X) = Y'$. Then, neglecting again a null set, Y' is a Borel complete section of E^Y_H, i.e., it meets every orbit of H. Let $y_1, y_2 \in Y'$ satisfy $y_1 E^Y_H y_2$. Say $y_1 = \bar{\rho}(x_1), y_2 = \bar{\rho}(x_2)$. Then $x_1 E^X_\Gamma x_2$, so let $\gamma \in \Gamma$ be such that $\gamma \cdot x_1 = x_2$. Then $\bar{\alpha}(\gamma, x_1) \cdot y_1 = y_2$, so $y_1 E^Y_{H'} y_2$. Thus in every H-orbit C, $Y' \cap C$ is contained in a single $E^Y_{H'}$-class. So if Y_0 is the H'-saturation of Y' and we assume that $H_0 \neq H_1$, towards a contradiction, so that $H' \neq H$, Y_0 and $Y \setminus Y_0$ both meet every E^Y_H-class, so they have positive measure. Now H' cannot be amenable, since then H_0 would be amenable and then Γ would be amenable (recall that $\Gamma/N \cong H_0$ and N is amenable). So by H'-ergodicity, one of $Y_0, Y \setminus Y_0$ is null, a contradiction. ⊣

Corollary 5.4. *Let H_0 be a countable group which contains no non-trivial normal amenable subgroups (e.g., a non-abelian free group or even a torsion-free hyperbolic group, which is not cyclic). Suppose Δ_0 is infinite amenable and $H_0 \times \Delta_0$ acts freely in a Borel way on X with invariant measure μ and the Δ_0-action is ergodic. Let H_1 be torsion-free hyperbolic (e.g., a free group), Δ_1 infinite amenable, and suppose $H_1 \times \Delta_1$ acts freely in a Borel way on Y with invariant measure ν and $H \times \Delta_1$ acts ergodically for any non-amenable subgroup $H \leq H_1$. If the action of $H_0 \times \Delta_0$ on X and of $H_1 \times \Delta_1$ on Y are stably orbit equivalent, then $H_0 \cong H_1$.*

We have seen in Chapter 1, **C**) that there are free measure preserving ergodic actions of F_2, F_3 which are SOE. In particular, by considering product actions, there are ergodic free measure preserving actions of $F_2 \times \mathbb{Z}, F_3 \times \mathbb{Z}$ which are SOE. However, the preceding corollary shows that if free measure preserving actions of $F_m \times \mathbb{Z}, F_n \times \mathbb{Z}$ are $\mathbb{Z}$-ergodic, and they are SOE, we must have $m = n$.

5B. Products of hyperbolic groups

Theorem 5.5. *Let $\Gamma_1, \Gamma_2, \ldots, \Gamma_n$ $(n \geq 1)$ be countable non-amenable groups, Δ a countable amenable group and let $\Gamma = \Gamma_1 \times \cdots \times \Gamma_n \times \Delta$ act freely in a Borel way on a standard Borel space X with invariant measure μ, so that the action of Δ is ergodic. Let $H_1, \ldots, H_n$ be torsion-free hyperbolic groups (e.g., free groups) and let $H = H_1 \times \cdots \times H_n$ act freely in a Borel way on Y. Then*

$$E^X_\Gamma \not\leq_B E^Y_H.$$

Proof. By induction on n.

Consider first the case $n = 1$. Suppose $E_\Gamma^X \leq_B E_H^Y$, where $\Gamma = \Gamma_1 \times \Delta, H = H_1$. Let $\rho : X \to Y$ be a Borel reduction of E_Γ^X to E_H^Y, and α the associated cocycle. Then, by 4.2, we have one of the following cases, in each of which we will derive a contradiction:

(i): α is equivalent to a cocycle with values in an amenable, thus cyclic-by-finite (see C4.1), subgroup of H. Then by changing ρ to another reduction, if necessary, and neglecting null sets, we can assume that α itself takes values in such a subgroup H_0 of H. Then clearly $E_\Gamma^X \leq_B E_{H_0}^Y$, so, since $E_{H_0}^Y$ is hyperfinite, E_Γ^X is hyperfinite, so Γ is amenable a contradiction.

(i)*: There is a hyperfinite equivalence relation F on a standard Borel space Z and a Borel homomorphism $\sigma : X \to Z$ of E_Γ^X to F with $\alpha(\gamma, x) = 1 \Leftrightarrow \sigma(x) = \sigma(\gamma \cdot x)$, μ-a.e. $(x), \forall \gamma \in \Gamma$. Without loss of generality we can assume this is true for all x. Then by 4.4, and neglecting μ-null sets, we conclude that there is a hyperfinite countable Borel equivalence relation $\hat{F}$ on $\hat{Z}$ such that ρ factors through $\hat{F}$, say $\rho = \rho_2 \circ \rho_1$, where ρ_1 is a Borel homomorphism of E_Γ^X to $\hat{F}$ and ρ_2 a Borel homomorphism of $\hat{F}$ to E_H^Y. It follows that ρ_1 is a reduction and so E_Γ^X is hyperfinite, thus Γ is amenable, a contradiction.

(ii)$'$: α is equivalent to a homomorphism $\pi : \Gamma \to H$ with $\pi(\Delta) = \{1\}$. Again we can assume, by changing ρ and by neglecting null sets, that actually $\alpha = \pi$. This implies that ρ is Δ-invariant, so that by the ergodicity of the Δ-action, it is constant a.e., a contradiction.

Now suppose $n > 1$ is arbitrary. Assume again ρ is a Borel reduction of E_Γ^X to E_H^Y, towards a contradiction, and let α be the associated cocycle. Let $\alpha_i = p_i \circ \alpha$, where p_i is the projection of H to H_i. We apply again 4.2 to each α_i. If (i) or (i)* of 4.2 holds for each i, then again by changing ρ to another reduction, if necessary, and neglecting null sets, we can assume that for each i, either α_i maps into an amenable subgroup of H_i or (i)* holds for α_i. Then by 4.4, neglecting null sets, ρ factors through some hyperfinite equivalence relation and we get a contradiction as in the case $n = 1$.

So for some $1 \leq i \leq n, \alpha_i$ satisfies (ii)$'$, i.e., $\alpha_i \sim \pi$, where $\pi : \Gamma \to H_i$ is a homomorphism, with $\pi(\Delta) = \{1\}$, and $\pi(\Gamma)$ not amenable. We claim then that there is a unique $1 \leq k \leq n$ such that $\pi(\Gamma_k) \neq \{1\}$. Indeed, there is at least one such k. Assume also $\pi(\Gamma_\ell) \neq \{1\}$, for some $\ell \neq k$. Fix $h \in \pi(\Gamma_k) \setminus \{1\}$, so that h has infinite order. Then for any $m \neq k, \pi(\Gamma_m) \subseteq C_{H_i}(h) =$ the centralizer of h in H_i. But $\langle h \rangle$ has finite index in $C_{H_i}(h)$, see [GdlH, p. 156, Théorème 34] (in case, H_i is free, of course $C_{H_i}(h)$ is cyclic), so $\pi(\Gamma_m)$ is amenable. Thus $\pi(\Gamma_m)$ is amenable, for each $m \neq k$, and similarly $\pi(\Gamma_m)$ is amenable, for each $m \neq \ell$. Since $k \neq \ell$, this shows that $\prod_{m=1}^n \pi(\Gamma_m) = \pi(\Gamma)$ is amenable, a contradiction.

We can assume without loss of generality that $k = i = 1$, so that $\pi(\Gamma_2) = \pi(\Gamma_3) = \cdots = \pi(\Gamma_n) = \{1\}$. Let $\Gamma_1' = \Gamma_1 \cap \ker(\pi)$, so that $\pi(\Gamma_1') = \{1\}$ as well. By changing ρ, if necessary, to another reduction, and neglecting null sets, we can assume that actually $\alpha_1 = \pi$. Notice then that ρ is a reduction of $E^X_{\Gamma_1' \times \Gamma_2 \times \cdots \times \Gamma_n \times \Delta}$ to $E^Y_{H_2 \times \cdots \times H_n}$. Indeed, if $x E^X_{\Gamma_1' \times \Gamma_2 \times \cdots \times \Gamma_n \times \Delta} y$, say $(\gamma_1', \gamma_2, \ldots, \gamma_n, \delta) \cdot x = y$ with $\gamma_1' \in \Gamma_1', \gamma_2 \in \Gamma_2, \ldots \gamma_n \in \Gamma_n, \delta \in \Delta$, then

$$(\alpha_1((\gamma_1', \gamma_2, \ldots, \gamma_n, \delta), x), \ldots, \alpha_n((\gamma_1', \gamma_2, \ldots, \gamma_n, \delta), x)) \cdot \rho(x) = \rho(y).$$

But

$$\alpha_1((\gamma_1', \ldots, \gamma_n, \delta), x) = \pi(\gamma_1', \ldots, \gamma_n, \delta) = 1,$$

so $\rho(x) E^Y_{H_2 \times \cdots \times H_n} \rho(y)$. Conversely, if $\rho(x) E^Y_{H_2 \times \cdots \times H_n} \rho(y)$, clearly $x E^X_\Gamma y$, so let $(\gamma_1, \ldots, \gamma_n, \delta) \cdot x = y$, with $\gamma_i \in \Gamma_i, \delta \in \Delta$. Then

$$(\alpha_1((\gamma_1, \ldots, \gamma_n, \delta), x), \ldots, \alpha_n((\gamma_1, \ldots, \gamma_n, \delta), x)) \in H_2 \times \cdots \times H_n,$$

i.e., $\alpha_1((\gamma_1, \ldots, \gamma_n, \delta), x) = \pi(\gamma_1) = 1$, so $\gamma_1 \in \Gamma_1'$ and $x F^X_{\Gamma_1' \times \Gamma_2 \times \cdots \times \Gamma_n \times \Delta} y$.

Put $\Gamma_2' = \Gamma_1' \times \Gamma_2, \Gamma_3' = \Gamma_3, \ldots, \Gamma_n' = \Gamma_n$. Then Γ_i' is not amenable, for $2 \leq i \leq n$ and $E^X_{\Gamma_2' \times \cdots \times \Gamma_n' \times \Delta} \leq_B E^Y_{H_2 \times \cdots \times H_n}$, violating the induction hypothesis. ⊣

Remark. For the case where the Γ_i's are free, non-abelian and the H_i's are free, and one of them is abelian, this result also follows from the results of Gaboriau [Ga01].

CHAPTER 6

Product Actions, I

In this and the next section, we will prove some results concerning relative non-reducibility of product actions, and use them in §8 to provide some applications concerning the relationship of the shift actions of product groups, whose factors are free or amenable groups, versus product actions of such groups. In general, if groups $\Gamma_1, \ldots, \Gamma_n$ act on spaces $X_1, \ldots, X_n$, resp., the *product action* of $\Gamma_1 \times \cdots \times \Gamma_n$ on $X_1 \times \cdots \times X_n$ is given by

$$(\gamma_1, \gamma_2, \ldots, \gamma_n) \cdot (x_1, \ldots, x_n) = (\gamma_1 \cdot x_1, \ldots, \gamma_n \cdot x_n).$$

If $E_1, \ldots, E_n$ are the equivalence relations induced by $\Gamma_1, \ldots, \Gamma_n$, resp., then the equivalence relation induced by the product action is $E_1 \times \cdots \times E_n$, where

$$(x_1, \ldots, x_n) E_1 \times \cdots \times E_n (y_1, \ldots, y_n) \Leftrightarrow \forall i \leq n (x_i E_i y_i).$$

Theorem 6.1. *Let Γ be a countable non-amenable group, which contains two infinite amenable subgroups $\Delta_1, \Delta_2 \leq \Gamma$ such that $\Delta_1 \subseteq N_\Gamma(\Delta_2), \Delta_2 \subseteq N_\Gamma(\Delta_1), \Gamma = N_\Gamma(\Delta_1) N_\Gamma(\Delta_2)$, where for any $\Delta \subseteq \Gamma, N_\Delta(\Gamma)$ is the normalizer of Δ in Γ. Suppose Γ acts freely in a Borel way on a standard Borel space X with invariant measure μ and assume that each infinite subgroup of Γ acts ergodically (e.g., the Γ-action is mixing). Let $E_1, \ldots, E_n$ be treeable countable Borel equivalence relations. Then*

$$E_\Gamma^X \not\leq_B E_1 \times \cdots \times E_n.$$

Corollary 6.2. *Let $\Gamma = \Gamma_1 \times \cdots \times \Gamma_m$, $m \geq 2$, be a countable non-amenable group, where at least two of the Γ_i's contain an infinite amenable subgroup. Then for any free mixing Borel action of Γ on X with invariant measure, E_Γ^X cannot be Borel reduced to a finite product of treeable countable Borel equivalence relations.*

Proof of Theorem 6.1. We will employ a variation of the ideas explained in Appendices B3, B4. We use the concept of the boundary of a tree, ∂T, as discussed in C1.

We will first introduce a few concepts that are needed in the proof.

Suppose the free group $F_2 = \langle a, b \rangle$ acts freely in a Borel way on a standard Borel space Y. Fix an orbit C of this action. Then define a tree T_C, with vertex set C, by

$$(y_1, y_2) \in T_C \Leftrightarrow \exists \gamma \in \{a^\pm, b^\pm\} (\gamma \cdot y_1 = y_2).$$

For any $y \in C$, the bijection $\varphi_y : C \to F_2$ given by $\varphi_y(z) =$ the unique $\gamma \in F_2$ with $\gamma^{-1} \cdot y = z$, clearly gives an isomorphism between T_C and the Cayley graph of F_2 (see C2), and induces a homeomorphism, also denoted by φ_y, of ∂T_C with ∂F_2. It also induces a Borel bijection of $[\partial T_C]^k$ with $[\partial F_2]^k$ and a homeomorphism of $\mathcal{M}(\partial T_C)$ and $\mathcal{M}(\partial F_2)$, all denoted by φ_y again. Note that if $\gamma \in F_2, \varphi_y(z) = \delta$ (so that $\delta \cdot z = y$), then $\varphi_{\gamma \cdot y}(z) = \gamma\delta$ (since $\gamma\delta \cdot z = \gamma \cdot y$). So $\varphi_{\gamma \cdot y}(z) = \gamma\varphi_y(z)$, thus, if e is in ∂T_C, then $\varphi_{\gamma\cdot y}(e) = \gamma \cdot \varphi_y(e)$, where in the right hand side we have the action of F_2 on ∂F_2.

Assume now $\pi : X \to Y$ is a Borel homomorphism of E^X_Γ to $E^Y_{F_2} = E$. As usual, we denote by $[Y]_E = F_2 \cdot y$, the E-class of y.

We will often consider below maps $x \mapsto S(x)$, with domain X and values $S(x)$, which are either finite subsets of $\partial T_{[\pi(x)]_E}$ or measures on $\partial T_{[\pi(x)]_E}$. Given a class of functions $\mathcal{F}$, whose domain is X and co-domain is a Polish space (e.g., $\mathcal{F}$ = Borel, universally measurable, μ-measurable, etc.), we want to define what it means for S to be in $\mathcal{F}$. Consider, for example, the case when $S(x)$ is a measure on $\partial T_{[\pi(x)]_E}$. We say then that S is in $\mathcal{F}$ if the map $S^* : X \to \mathcal{M}(\partial F_2)$, defined by

$$S^*(x) = \varphi_{\pi(x)}(S(x)),$$

is in the class $\mathcal{F}$.

For the remainder of the proof, it will be convenient to assume the Continuum Hypothesis (CH). This is harmless, as explained in the proof of 4.4.

We will derive the theorem from the following lemma:

Lemma 6.3. *Let Γ, X, μ be as in the hypothesis of Theorem 6.1, and let F_2 act freely in a Borel way on a standard Borel space Y with induced equivalence relation $E^Y_{F_2} = E$, and let $C \mapsto T_C$, for $C \in Y/E$, be the corresponding trees. Let $\pi : X \to Y$ be a Borel homomorphism of E^X_Γ to E, and assume that it does not map μ-a.e. into a single E-class. Then there is a universally measurable map $x \mapsto S(x)$ such that:*

(i) $S(x) \subseteq \partial T_{[\pi(x)]_E}$, μ-a.e. (x).

(ii) $0 < |S(x)| \leq 2$, μ-a.e. (x).

(iii) $\forall \gamma \in \Gamma(S(\gamma \cdot x) = S(x))$, μ-a.e. (x), i.e., S is Γ-invariant a.e.

Granting this lemma, we can complete the proof of 6.1 as follows:

If 6.1 fails, we can assume that $E^X_\Gamma \leq_B E_1 \times \cdots \times E_n$, and n is least so that for some action of Γ on some X', μ' satisfying the hypotheses of 6.1, $E^{X'}_\Gamma$ can be Borel reduced to a product of n treeable equivalence relations. By 1.1, we can also assume that E_i $(1 \leq i \leq n)$ is induced by a free Borel action of F_2 on a standard Borel space Y_i. Fix then a Borel reduction $\rho : X \to Y_1 \times \cdots \times Y_n$ of E^X_Γ to $E_1 \times \cdots \times E_n$.

Write $\rho = (\pi_1, \ldots, \pi_n)$, where $\pi_i = p_i \circ \rho$, with $p_i : Y_1 \times \cdots \times Y_n \to Y_i$ the ith projection function. We will apply Lemma 6.3 to $\Gamma, X, \mu, E_i, \pi_i$.

If $\mu(\pi_i^{-1}(C)) = 1$ for some E_i-class C, then letting $X' = \pi_i^{-1}(C)$, which is Γ-invariant, we see that $\pi' = (\pi_1, \ldots, \pi_{i-1}, \pi_{i+1}, \ldots, \pi_n)$ is such that

$$xE^{X'}_\Gamma y \Leftrightarrow \pi'(x)(E_1 \times \cdots \times E_{i-1} \times E_{i+1} \times \cdots \times E_n)\pi'(y),$$

contradicting the minimality of n. (If $n = 1$, then $\rho : X \to Y_1$ maps into a single E_1-class, μ-a.e., contradicting the fact that ρ is a reduction. Thus $n \geq 2$, if $\mu(\pi_i^{-1}(C)) = 1$ for some i and some E_i-class C.)

It follows that $\mu(\pi_i^{-1}(C)) = 0$, for all $i = 1, \ldots, n$, and all E_i-classes C. Apply then 6.3 to each $\pi_i, i = 1, \ldots, n$ to find a universally measurable map $x \mapsto S_i(x)$ such that for μ-a.e. (x):

(i) $S_i(x) \subseteq \partial T_{[\pi_i(x)]_{E_i}}$.

(ii) $0 < |S_i(x)| \leq 2$.

(iii) $S_i(x)$ is Γ-invariant.

By restricting to a Γ-invariant set of measure 1, we can assume that each S_i is actually defined everywhere and Borel.

Let $[\rho(X)]_{E_1\times\cdots\times E_n} = A \subseteq Y_1 \times \cdots \times Y_n$. We will argue that $(E_1 \times \cdots \times E_n)|A$ is measure-amenable (see [JKL, 8.3]), so since $E_\Gamma^X \leq_B (E_1 \times \cdots \times E_n)|A$, E_Γ^X is measure-amenable, so Γ is amenable, a contradiction.

We will consider the case where $n = 2, |S_1(x)| = 1, |S_2(x)| = 2$. The general case is only notationally more complicated.

Consider an $(E_1 \times E_2)$-class C contained in A. Then $C = C_1 \times C_2$, where $C_1 \in Y_1/E_1, C_2 \in Y_2/E_2$, and, since $|S_1(x)| = 1, |S_2(x)| = 2$, we can assign to C an end $e_C^1 \in \partial T_{C_1}$ and a pair $\{d_C^2, e_C^2\} \subseteq \partial T_{C_2}$. We will use this to define means φ_1^C on C_1 and φ_2^C on C_2. We will then take $\varphi_C = \varphi_1^C \otimes \varphi_2^C$, the Fubini product of φ_1^C, φ_2^C. It will be straightforward to verify that $C \mapsto \varphi_C$ is universally measurable, in the sense of [JKL, 2.7]. Thus $(E_1 \times E_2)|A$ is measure amenable, as claimed.

To define φ_1^C, fix any $\alpha \in C_1$ and let $\alpha_0 = \alpha, \alpha_1, \alpha_2 \ldots$ be the geodesic from α to e_C^1. Put $T^n(\alpha) = \alpha_n$. Define then for any $f \in \ell_\infty(C_1)$,

$$\varphi_\alpha^1(f) = \int \frac{f(T^0(\alpha)) + \cdots + f(T^n(\alpha))}{n+1} d\theta(n),$$

where θ is a universally measurable shift invariant mean on $\mathbb{N}$ (see [JKL, 2.1]–we are using CH here). Then it is easy to see that φ_α^1 is independent of α, so we put

$$\varphi_1^C = \varphi_\alpha^1, \text{ for any } \alpha \in C_1.$$

To define φ_2^C, note that by the usual arguments (see [JKL, 3.19] and [Ke01, 8.2]), we can find equivalence relations $\{E_{n,C}^2\}$ on C_2^*, where $C_2^* = [d_C^2, e_C^2]$ is the line determined by $\{d_C^2, e_C^2\}$, depending in a Borel way on any representative $(x_1, x_2) \in C$, such that

(a) $E_{0,C}^2 \subseteq E_{1,C}^2 \subseteq E_{2,C}^2 \subseteq \ldots$,

(b) $\bigcup_n E_{n,C}^2 = C_2^* \times C_2^*$,

(c) Each $E_{n,C}^2$ is finite (i.e., has finite classes).

For each $E_{n,C}^2$ class D now define the mean $\varphi_{n,D}^2$ by

$$\varphi_{n,D}^2(f) = \frac{\sum_{\beta\in D} f(\beta)}{|D|},$$

for $f \in \ell_\infty(D) = \mathbb{R}^D$ (as D is finite). Then put for any $\beta \in C_2^*, f \in \ell_\infty(C_2)$,

$$\varphi_\beta^2(f) = \int \varphi_{n,[\beta]_{E_{n,C}^2}}^2 (f|[\beta]_{E_{n,C}^2}) d\theta(n).$$

Again it is easy to check that φ_β^2 is independent of $\beta \in C_2^*$, so put

$$\varphi_2^C = \varphi_\beta^2, \text{ for any } \beta \in C_2^*.$$

It only remains to give the

Proof of Lemma 6.3.

Fix $\Delta \subseteq \Gamma$ infinite amenable.

Lemma 6.4. *There is a map $x \mapsto S^0(x)$, defined on X, such that*

(i) $S^0(x) \subseteq \partial T_{[\pi(x)]_E}$,

(ii) $0 < |S^0(x)| \leq 2$,

(iii) S^0 is Δ-invariant.

(iv) S^0 is universally measurable.

Proof. By B3.2 applied to the Δ-action on $X, H = F_2, K = \partial F_2, \alpha$ the (strict) cocycle associated to π (restricted to $\Delta \times X$), we see that there is a universally measurable map $x \mapsto \nu_x$ from X to $\mathcal{M}(\partial F_2)$, with $\alpha(\delta, x) \cdot \nu_x = \nu_{\delta \cdot x}$, for $x \in X, \delta \in \Delta$. Let

$$\nu'_x = \varphi^{-1}_{\pi(x)}(\nu_x) \in \mathcal{M}(\partial T_{[\pi(x)]_E}).$$

Then it is easy to verify that $x \mapsto \nu'_x$ is universally measurable and Δ-invariant. If $\nu'_x \in \mathcal{M}_{\leq 2}(\partial T_{[\pi(x)]_E})$, let $S^0(x) = \text{supp}(\nu'_x)$. If $\nu'_x \in \mathcal{M}_3(\partial T_{[\pi(x)]_E})$, then, by the argument in C2.3, $(\nu'_x)^3([\partial T_{[\pi(x)]_E}]^3) > 0$, and if $\varphi(\{e_1, e_2, e_3\}) = [e_1, e_2, e_3]$ (see the notation in C1), for $\{e_1, e_2, e_3\} \in [\partial T_{[\pi(x)]_E}]^3$, then the image, ν''_x, of the normalized restriction of $(\nu'_x)^3$ to $[\partial T_{[\pi(x)]_E}]^3$, is a measure on $[\pi(x)]_E$, so let A_x be the finite set of elements of $[\pi(x)]_E$, which have maximum measure. Let $\eta : [F_2]^{<\infty} \to F_2$ be an F_2-map (see the proof of C2.3) and let $S^*(x) = \varphi^{-1}_{\pi(x)}(\eta(\varphi_{\pi(x)}(A_x))) \in [\pi(x)]_E$. Put $S^0(x) = \{b(S^*(x))\}$, where b is a Borel map that assigns to each $y \in Y$ an end $b(y) \in \partial T_{[y]_E}$.

It is clear that this S^0 works. ⊣

Lemma 6.5. *Suppose $x \mapsto S(x)$ satisfies:*
(i) $S(x) \subseteq \partial T_{[\pi(x)]_E}$, μ-a.e. (x),
(ii) $S(x) \neq \emptyset$ and is finite, μ-a.e. (x),
(iii) S is Δ-invariant, μ-a.e.,
(iv) S is universally measurable.
Then $|S(x)| \leq 2$, μ-a.e. (x).

Proof. Let $A = \{x : |S(x)| > 2\}$. Then if $\mu(A) > 0$, find a Borel set $B \subseteq A$ which is Δ-invariant and $\mu(B) > 0$. Then by the argument in the preceding lemma, we can find, shrinking B a bit if necessary, a Borel function g on B which is Δ-invariant and $g(x) \in [\pi(x)]_E$. Then, since the Δ-action is ergodic, $g(x) = y_0$, μ-a.e. (x), for some $y_0 \in Y$. Thus $y_0 \in [\pi(x)]_E$, μ-a.e. (x), so $\pi(x) \in [y_0]_E = C$, μ-a.e. (x), a contradiction. ⊣

Denote by $\mathcal{S}_\Delta$ the set of all functions $x \mapsto S(x)$ satisfying:
(i) $S(x) \subseteq \partial T_{[\pi(x)]_E}$, μ-a.e. (x),
(ii) $0 < |S(x)| \leq 2$, μ-a.e. (x),
(iii) S is Δ-invariant, μ-a.e.,
(iv) S is universally measurable.
Define a partial (pre-)order $\preceq$ on $\mathcal{S}_\Delta$ by

$$S \preceq T \text{ iff } S(x) \subseteq T(x), \ \mu\text{-a.e. } (x).$$

Let also for $S \in \mathcal{S}_\Delta$,

$$D(S) = \{x : |S(x)| = 2\}.$$

The following is an analog of B4.1 and can be proved by the same argument, but we prefer to present the proof in a different way for further use.

Lemma 6.6. *The partial order $(\mathcal{S}_\Delta, \preceq)$ has a maximum element, denoted by S_Δ. (This is of course unique a.e.)*

Proof. Assume not, towards a contradiction. Define by transfinite induction functions $\{S_\alpha\}_{\alpha<\omega_1}$ in $\mathcal{S}_\Delta$ as follows: $S_0 = S^0$ is given by Lemma 6.4. Assume now S_α is defined. It is not a maximum, so there is $S \in \mathcal{S}_\Delta$ such that $S(x) \not\subseteq S_\alpha(x)$ on a set of positive measure. Put $S_{\alpha+1} = S_\alpha \cup S$. By Lemma 6.5, $S_{\alpha+1} \in \mathcal{S}_\Delta$. Clearly $S_{\alpha+1} \succeq S_\alpha$. Put

$$D_\alpha = \{x : |S_\alpha(x)| = 2\},$$
$$D_{\alpha+1} = \{x : |S_{\alpha+1}(x)| = 2\}.$$

Since $|S_{\alpha+1}(x)| \leq 2$ μ-a.e. (x) and $S_\alpha \preceq S_{\alpha+1}$, clearly $D_\alpha \subseteq D_{\alpha+1}$ μ-a.e. (Here we let $A \subseteq B$ μ-a.e., stand for $\mu(A \setminus B) = 0$, and $A \subsetneqq B$ μ-a.e., for $\mu(A \setminus B) > 0$.) Consider any x with $S(x) \not\subseteq S_\alpha(x)$. Then $|S_{\alpha+1}(x)| \geq 2$, so $x \in D_{\alpha+1}$, and also $x \notin D_\alpha$ μ-a.e. (x), so $\mu(D_{\alpha+1} \setminus D_\alpha) > 0$. Thus we have

$$D_\alpha \subsetneqq D_{\alpha+1} \ \mu\text{-a.e.}$$

Finally assume S_α have been defined for all $\alpha < \lambda$, λ limit, with $S_\alpha \preceq S_\beta$ and $D_\alpha \subsetneqq D_\beta$ μ-a.e. for all $\alpha < \beta < \lambda$. Put $S_\alpha = \bigcup_{\alpha<\lambda} S_\alpha$. We claim that $S_\alpha \in \mathcal{S}_\Delta$. Then clearly $S_\alpha \preceq S_\lambda, \forall \alpha < \lambda$, and $D_\alpha \subsetneqq D_\lambda$ μ-a.e., $\forall \alpha < \lambda$.

Put $D = \bigcup_{\alpha<\lambda} D_\alpha$. If $x \in D, |S_\lambda(x)| = 2$, since if $x \in D_\alpha$, then $S_\alpha(x) = S_\beta(x), \forall \alpha \leq \beta < \lambda$, μ-a.e (x). If $x \notin D$, then $|S_\alpha(x)| = 1, \forall \alpha < \lambda$, and $S_\alpha(x) \subseteq S_\beta(x), \forall \alpha < \beta < \lambda$, so $S_\alpha(x) = S_\beta(x) = S_\lambda(x)$, and $|S_\lambda(x)| = 1$.

Since we have $D_\alpha \subsetneqq D_\beta$ μ-a.e., $\forall \alpha < \beta < \omega_1$, the sets $\{D_{\alpha+1} \setminus D_\alpha\}_{\alpha<\omega_1}$ clearly violate the countable chain condition, a contradiction. ⊣

Lemma 6.7. *S_Δ is $N_\Gamma(\Delta)$-invariant, μ-a.e.*

Proof. Fix $\gamma \in N_\Gamma(\Delta)$. Put

$$S_\gamma(x) = S_\Delta(\gamma \cdot x).$$

Then for $\delta \in \Delta$,

$$\begin{aligned} S_\gamma(\delta \cdot x) &= S_\Delta(\gamma\delta \cdot x) = S_\Delta(\delta'\gamma \cdot x), \text{ for some } \delta' \in \Delta, \\ &= S_\Delta(\delta' \cdot (\gamma \cdot x)) = S_\Delta(\gamma \cdot x) \\ &= S_\gamma(x), \ \mu\text{-a.e. } (x). \end{aligned}$$

So S_γ is Δ-invariant, thus $S_\gamma \preceq S_\Delta$. Also

$$\begin{aligned} x \in D(S_\gamma) &\Leftrightarrow |S_\gamma(x)| = 2 \\ &\Leftrightarrow |S_\Delta(\gamma \cdot x)| = 2 \\ &\Leftrightarrow \gamma \cdot x \in D(S_\Delta). \end{aligned}$$

So $\mu(D(S_\gamma)) = \mu(D(S_\Delta))$ and since $S_\gamma \preceq S_\Delta$, thus $D(S_\gamma) \subseteq D(S_\Delta)$, μ-a.e., we have $D(S_\gamma) = D(S_\Delta)$, μ-a.e., so $S_\gamma = S_\Delta$, μ-a.e. Thus $S_\Delta(\gamma \cdot x) = S_\Delta(x)$, μ-a.e. (x), i.e., S_Δ is γ-invariant a.e., and so S_Δ is $N_\Gamma(\Delta)$-invariant a.e. ⊣

We complete now the proof of Lemma 6.3:

Fix two infinite amenable subgroups $\Delta_1, \Delta_2 \leq \Gamma$ with $\Delta_1 \subseteq N_\Gamma(\Delta_2), \Delta_2 \subseteq N_\Gamma(\Delta_1)$ and $N_\Gamma(\Delta_1)N_\Gamma(\Delta_2) = \Gamma$. Consider the two functions $S_{\Delta_1}, S_{\Delta_2}$. As $\Delta_2 \subseteq N_\Gamma(\Delta_1), S_{\Delta_1}$ is also Δ_2-invariant μ-a.e. and so $S_{\Delta_1} \subseteq S_{\Delta_2}$ μ-a.e. Similarly $S_{\Delta_2} \subseteq S_{\Delta_1}$ μ-a.e., so $S_{\Delta_1} = S_{\Delta_2}$ μ-a.e. Thus letting $S = S_{\Delta_1}$, clearly S is both $N_\Gamma(\Delta_1)$ and $N_\Gamma(\Delta_2)$-invariant μ-a.e., so it is Γ-invariant μ-a.e. ⊣

Remark. It appears that, using the results in [Ad94], one could also extend 6.1, so that it also applies to $E_1, \ldots, E_n$ induced by free Borel actions of hyperbolic groups, but we have not verified the details.

CHAPTER 7

Product Actions, II

We will employ here also some concepts and results from Gaboriau [Ga01]. In particular, we recall that the *approximate ergodic dimension* of a countable group Γ is the smallest n such that for some free Borel action of Γ with invariant measure on some X, E_Γ^X can be written as $\bigcup_m E_m$, with $\{E_m\}$ an increasing sequence of Borel equivalence relations each of which can be Borel reduced to a $\mathcal{K}_n$-structurable countable Borel equivalence relation, in the sense of Appendix D (here $\mathcal{K}_n$ = the class of n-dimensional contractible simplicial complexes). If $\beta_n(\Gamma)$ denotes the nth l^2-Betti number of Γ (see [Ga01, 1.2]), and Γ has approximate ergodic dimension $< n$, then $\beta_n(\Gamma) = 0$ (see [Ga01, 5.13]). In particular, as $\beta_n(F_2^n) = 1 \neq 0$ for $n \geq 1$ (see [Ga01, 1.5, 1.6]), F_2^n has approximate ergodic dimension equal to n.

Theorem 7.1. *Let $n \geq 1$. Let $\Gamma = \Gamma_1 \times \Gamma_2$, where Γ_1 has approximate ergodic dimension $\geq n$ and Γ_2 contains an infinite amenable subgroup. Assume Γ_1 acts freely in a Borel way on a standard Borel space X_1 with invariant measure μ_1 and Γ_2 acts freely in a Borel way on a standard Borel space X_2 with invariant measure μ_2. Consider the product action of $\Gamma = \Gamma_1 \times \Gamma_2$ on $X = X_1 \times X_2$ (which has invariant measure $\mu = \mu_1 \times \mu_2$). Then if $E_1, \dots, E_n$ are treeable countable Borel equivalence relations,*

$$E_\Gamma^X \not\leq_B E_1 \times \cdots \times E_n.$$

Corollary 7.2. *Let $n \geq 1$. Suppose F_2^n acts freely in a Borel way on X_1 with invariant measure μ_1 and $\mathbb{Z}$ acts freely in a Borel way on X_2 with invariant measure μ_2. Then if $F_2^n \times \mathbb{Z}$ acts by the product action on $X_1 \times X_2$,*

$$E_{F_2^n \times \mathbb{Z}}^{X_1 \times X_2} \not\leq_B E_1 \times \cdots \times E_n,$$

for any treeable Borel equivalence relations $E_1, \dots, E_n$.

Proof of Theorem 7.1. We can assume that E_i is induced by a free Borel action of F_2 on a standard Borel space Y_i. Assume then that $\rho : X \to Y = Y_1 \times \cdots \times Y_n$ is a Borel reduction of E_Γ^X to $E_1 \times \cdots \times E_n$, towards a contradiction. As in the proof of 6.1, we will make the harmless assumption that the Continuum Hypothesis (CH) holds. Let $p_i : Y \to Y_i$ be the ith projection and put $\pi_i = p_i \cdot \rho$. Finally we fix an infinite amenable subgroup $\Delta \subseteq \Gamma_2$.

Lemma 7.3. *There are maps $x \mapsto S_i^0(x), i = 1, \dots, n$, such that*

(i) $S_i^0(x) \subseteq \partial T_{[\pi_i(x)]_{E_i}}$.

(ii) $0 < |S_i^0(x)| \leq 2$.

(iii) S_i^0 is Δ-invariant.

(iv) S_i^0 is universally measurable.

Proof. As in Lemma 6.4. ⊣

For each Δ-invariant Borel set $A \subseteq X$, and $1 \leq i \leq n$, denote by $\mathcal{S}_i^A$ the set of functions $x \mapsto S(x)$ defined on A and satisfying:

(i) $S(x) \subseteq \partial T_{[\pi_i(x)]_{E_i}}$, μ-a.e. (x).

(ii) $0 < |S(x)| \leq 2$, μ-a.e. (x).

(iii) S is Δ-invariant, μ-a.e.

(iv) S is universally measurable.

As usual, we define the partial order $\preceq$ on $\mathcal{S}_i^A$ by

$$S \preceq T \Leftrightarrow S(x) \subseteq T(x),\ \mu\text{-a.e. } (x).$$

We call the set A *i-nice* if $\mathcal{S}_i^A$ has a maximum element in $\preceq$.

Lemma 7.4. *For every Δ-invariant Borel set A of positive measure, there is a Δ-invariant Borel set $B \subseteq A$ of positive measure and $i \in \{1, \dots, n\}$ such that B is i-nice.*

Proof. Consider the functions $(S_1, \dots, S_n) = (S_1^0, \dots, S_n^0)$ given in Lemma 7.3. Start with $S_1|A$ and define a transfinite sequence $\{T_\alpha\}_{\alpha<\beta_1}$ (β_1 some countable ordinal to be determined later) of functions in $\mathcal{S}_1^A$, as follows:

$$T_0 = S_1|A$$

Suppose T_α has been defined. If T_α is maximum in $\mathcal{S}_1^A$, we stop the construction at α, i.e., we let $\beta_1 = \alpha + 1$, and we are done, as A itself is 1-nice. Else there is $S \in \mathcal{S}_1^A$ so that $S(x) \not\subseteq T_\alpha(x)$ on a set of positive measure. Put $\tilde{T}_{\alpha+1} = S \cup T_\alpha$. Clearly $\tilde{T}_{\alpha+1}$ satisfies (i), (iii) of Lemma 7.3 a.e. on A, and also (iv). If it also satisfies (ii) a.e. on A, then clearly $\tilde{T}_{\alpha+1} \in \mathcal{S}_1^A$ and we put $T_{\alpha+1} = \tilde{T}_{\alpha+1}$. Note that if we let $D_\alpha = \{x \in A : |T_\alpha(x)| = 2\}$, then $D_\alpha \subsetneqq D_{\alpha+1}$ μ-a.e. Otherwise, $\tilde{T}_{\alpha+1}$ fails to satisfy (ii) a.e. on A, so there is a Δ-invariant Borel set $A_1 \subseteq A$ of positive measure and a function U_1 on A_1 satisfying (i), (iii). a.e. on A_1, satisfying also (iv), and finally such that $3 \leq |U_1(x)| \leq 4$ a.e. on A_1. If this happens, we stop the construction at α (i.e., we let $\beta_1 = \alpha + 1$) and we say that we are in the *bad$_1$-case.*

Finally, if T_α has been defined for all $\alpha < \lambda$, λ a limit ordinal, so that $T_\alpha \preceq T_\beta$ and $D_\alpha \subsetneqq D_\beta$, μ-a.e., for all $\alpha < \beta < \lambda$, we take $T_\lambda = \bigcup_{\alpha<\lambda} T_\alpha$, so that $T_\lambda \in \mathcal{S}_1^A$ and $T_\alpha \preceq T_\lambda, D_\alpha \subsetneqq D_\lambda$, μ-a.e., $\forall \alpha < \lambda$.

Since this process cannot, by the countable chain condition, go on for all $\alpha < \omega_1$, it follows that it must stop, either because some T_α is maximum, in which case we are done, or else because we hit a bad$_1$-case with witness (A_1, U_1).

Now repeat the same procedure on A_1 starting with $S_2|A_1$ (and working within $\mathcal{S}_2^{A_1}$), etc. Proceeding this way for m steps if necessary, we see that either we are done, or else we can find a set $C = A_n \subseteq A_{n-1} \subseteq \cdots \subseteq A_1 \subseteq A$ of positive measure which is Δ-invariant, and we can find functions $U_1, \dots, U_n$ on C which satisfy (i), (iii) of Lemma 7.3 a.e. on C, satisfy (iv), and also $3 \leq |U_i(x)| \leq 4$ a.e. on C, $i = 1, \dots, n$. Shrinking C a bit, if necessary, we can find, as in the proof of Lemma 6.5, a Borel function $g : C \to Y$ such that $g(x) \in [\rho(x)]_{E_1 \times \cdots \times E_n}, \forall x \in C$. Since Δ acts freely on C with invariant measure, $E_{\Delta_1}^{A_1}$ is not tame, so there is, by a version of the Glimm-Effros Dichotomy (see, e.g., [DJK, 3.4]), or the ergodic decomposition theorem, a Δ-ergodic, non-atomic measure on C. As in the proof of 6.5, this leads to contradiction. ⊣

Lemma 7.5. *For each $i = 1, \ldots n$, there is a maximum i-nice Borel set, i.e., there is a Borel set A_i which is i-nice, and for any Borel set A which is i-nice, we have $A \subseteq A_i$ μ-a.e.*

Proof. By a straightforward exhaustion argument. ⊣

By Lemma 7.4, we have

$$A_1 \cup A_2 \cup \cdots \cup A_n = X, \ \mu\text{-a.e.}.$$

Denote also by M_i the maximum element of $\mathcal{S}_i^{A_i}, i = 1, \ldots, n$.

Lemma 7.6. *A_i, M_i is $\Gamma_1 \times \Delta$-invariant, μ-a.e.*

Proof. Fix $\gamma_1 \in \Gamma_1$. Consider $\gamma_1 \cdot A_i$. It is clearly Δ-invariant. Define on it

$$f_{\gamma_1}(x) = M_i((\gamma_1)^{-1} \cdot x).$$

Then f_{γ_1} is the maximum element of $\mathcal{S}_i^{\gamma_1 \cdot A_i}$, so $\gamma_i \cdot A_i$ is i-nice, thus $\gamma_1 \cdot A_i \subseteq A_i$ μ-a.e. It follows that A_i is Γ_i-invariant and thus $\Gamma_1 \times \Delta$-invariant μ-a.e. Similarly for M_i. ⊣

Since $A_1 \cup A_2 \cup \cdots \cup A_n = X$ μ-a.e., one of the A_i's has positive measure, say it is A_1. Since $A_1 \subseteq X_1 \times X_2$ has positive measure, by Fubini we can find $x_2 \in X_2$ such that if $(A_1)^{x_2} = \{x_1 : (x_1, x_2) \in A_1\}$, then $\mu_1((A_1)^{x_2}) > 0$. Put $A' = (A_1)^{x_2}$. Clearly, A' is Γ_1-invariant, since A_1 is $\Gamma_1 \times \Delta$-invariant. For $x_1 \in A'$, let

$$\rho'(x_1) = \rho(x_1, x_2).$$

Then for $x_1, y_1 \in A'$:

$$\begin{aligned} x_1 E_{\Gamma_1}^{A'} y_1 &\Leftrightarrow (x_1, x_2) E_{\Gamma_1 \times \Gamma_2}^{X} (y_1, x_2) \\ &\Leftrightarrow \rho(x_1, x_2)(E_1 \times \cdots \times E_n)\rho(y_1, x_2) \\ &\Leftrightarrow \rho'(x_1)(E_1 \times \cdots \times E_n)\rho'(y_2), \end{aligned}$$

i.e., ρ' reduces $E_{\Gamma_1}^{A'}$ to $E_1 \times \cdots \times E_n$. Also if, for $x_1 \in A'$, we put

$$M'(x_1) = M_1(x_1, x_2),$$

we have (letting $\rho' = (\pi_1', \ldots, \pi_n')$):
(i) $M'(x_1) \subseteq \partial T_{[\pi_1'(x_1)]_{E_1}}$, a.e.
(ii) $0 < |M'(x_1)| \leq 2$, a.e.
(iii) M' is Γ_1-invariant a.e.
(iv) M' is universally measurable.
We will now derive a contradiction from this.

By shrinking A' a bit, we can assume that M' is actually Borel and (i)-(iii) hold for every $x_1 \in A'$. Put

$$Z = [\rho'(A')]_{E_1 \times \cdots \times E_n}.$$

This is a Borel $(E_1 \times \cdots \times E_n)$-invariant subset of $Y = Y_1 \times \cdots \times Y_n$. Put $Y' = Y_2 \times \cdots \times Y_n, E' = E_2 \times \cdots \times E_n$. Then every $(E_1 \times \cdots \times E_n) = (E_1 \times E')$-class is of the form $C = C_1 \times C'$, with C_1 an E_1-class and C' an E'-class. Also E' is $\mathcal{K}_{n-1}$-structurable, where $\mathcal{K}_{n-1}$ = the class of simplicial complexes of dimension $\leq n-1$ which are contractible (see the last paragraph of Appendix D).

Now fix an $(E_1 \times E')$-class $C = C_1 \times C'$ contained in Z, so that $C = [\rho'(x_1)]_{E_1 \times E'}$, for some $x_1 \in A'$. Consider two cases, according to whether $|M'(x_1)| = 1$ or 2.

Case 1. $|M'(x_1)| = 1$ (note that this only depends on the $E_{\Gamma_1}^{A'}$-class of x_1).

For $y_1 \in C_1$, let $y_1, y_2 \ldots$ be the geodesic from y_1 to $M'(x_1)$ (we view here $M'(x_1)$ as an element of $\partial T_{[\pi_1(x_1)]_{E_1}} = \partial T_{C_1}$, as opposed to a singleton). Consider the tail equivalence relation E_t on $Y_1^{\mathbb{N}}$

$$(u_1, u_2, \ldots) E_t (v_1, v_2, \ldots)$$
$$\Leftrightarrow \exists n \exists m \forall k (u_{n+k} = v_{m+k}).$$

This is hypertame, i.e., the union of an increasing sequence of tame Borel equivalence relations (see [DJK, §8]), so let $E_t = \bigcup_m R_m$, where (R_m) is an increasing sequence of tame Borel equivalence relations on $Y_1^{\mathbb{N}}$. Let $f_m : Y_1^{\mathbb{N}} \to \mathbb{R}$ be such that $u R_m v \Leftrightarrow f_m(u) = f_m(v)$. For $(y_1, y') \in C$, put

$$g_m(y_1, y') = (f_m(y_1, y_2, \ldots), 1) \in \mathbb{R} \times \{1, 2\}.$$

Note that for $(y_1, y) \in C, (z_1, z') \in C$:

(a) $g_m(y_1, y') = g_m(z_1, z') \Rightarrow g_{m+1}(y_1, y') = g_{m+1}(z_1, z')$.

Case 2. $|M'(x_1)| = 2$.

Then $M'(x_1)$ is a pair of ends in ∂T_{C_1}, so it defines a unique line $L(x_1)$ in T_{C_1}. $L(x_1)$ is Γ_1-invariant as well. By the usual arguments (see [BK, proof of Lemma 4.5.3] and [Ke01, proof of 8.2]) we can define, in a Γ_1-invariant way, finite subequivalence relations $E'_m(x_1)$ on $L(x_1)$ such that $E'_m(x_1) \subseteq E'_{m+1}(x_1)$ and $\bigcup_m E'_m(x_1) = L(x') \times L(x')$. For $y_1 \in C_1$, let $v(y_1)$ be the vertex in $L(x_1)$ of least distance from y_1. Assume, without loss of generality, that Y_1 is a Borel subset of $\mathbb{R}$, and put $\bar{g}_m(y_1) =$ the least element of $[v(y_1)]_{E'_m(x_1)}$. Finally, for $(y_1, y') \in C$, put

$$g_m(y_1, y') = (\bar{g}_m(y_1), 2) \in \mathbb{R} \times \{1, 2\}.$$

Again note that for $(y_1, y'), (z_1, z') \in C$

(a) $g_m(y_1, y') = g_m(z_1, z') \Rightarrow g_{m+1}(y_1, y') = g_{m+1}(z_1, z')$.

This completes the definition of the functions g_m on Z. We next note the following property:

(b) If C, D are two $(E_1 \times E')$-classes contained in Z, and $(y_1, y') \in C, (z_1, z') \in D$ are such that $g_m(y_1, y') = g_m(z_1, z')$, then $C_1 = D_1$.

To see this, notice that, since $g_m(y_1, y') = g_m(z_1, z'), C, D$ are both in Case 1 or both in Case 2. If they are in Case 1, then $f_m(y_1, y_2, \ldots) = f_m(z_1, z_2, \ldots)$, so $(y_1, y_2, \ldots) R_m (z_1, z_2, \ldots)$, thus $y E_t x$, so $\exists k, \ell$ with $y_k = z_\ell$ and thus $C_1 = D_1$. If they are in Case 2, then $\bar{g}_m(y_1) = \bar{g}_m(z_1)$, so again $C_1 = D_1$.

Now define a subequivalence relation Q_m of $(E_1 \times E')|Z$ by

$$(y_1, y') Q_m (z_1, z') \Leftrightarrow (y_1, y')(E_1 \times E')(z_1, z') \ \& \ g_m(y_1, y') = g_m(z_1, z').$$

Then $Q_m \subseteq Q_{m+1}$ and $\bigcup_m Q_m = (E_1 \times E')|Z$.

Claim. Q_m is Borel reducible to a $\mathcal{K}_{n-1}$-structurable equivalence relation.

Granting this, let $Q'_m = (\rho')^{-1}(Q_m)$. Then $Q'_m \subseteq Q'_{m+1}, \bigcup_m Q'_m = E_{\Gamma_1}^{A'}$ and Q'_m is Borel reducible to a $\mathcal{K}_{n-1}$-structurable equivalence relation, and thus the approximate ergodic dimension of Γ_1 is $< n$, a contradiction.

So it only remains to verify the Claim.

Proof of Claim. By Appendix D.1 it is enough to verify that Q_m admits a tame $\mathcal{K}_{n-1}$-structured Q_m-space.

Consider the following fiber space U over Z:

$$(y_1, y', v, z') \in U \Leftrightarrow (y_1, y') \in Z \ \& \ v = g_m(y_1, y') \ \& \ z'E'y',$$

with projection function

$$(y_1, y', v, w) \mapsto (y_1, y').$$

Fixing $(y_1, y') \in Z$ and $v = g_m(y_1, y')$, use the map $z' \in [y']_{E'} \mapsto (y_1, y', v, z')$ to carry over the contractible simplicial complex of dimension $\leq n-1$ from $[y']_{E'}$ to the fiber above (y_1, y'). If $(y_1, y')Q_m(z_1, z')$ and $v = g_m(y_1, y') = g_m(z_1, z')$, then the action of Q_m on U is defined by

$$((y_1, y'), (z_1, z')) \cdot (z_1, z', v, t') = (y_1, y', v, t').$$

This is clearly an isomorphism of the corresponding complexes. It only remains to verify that the corresponding equivalence relation of this action is tame. Call this equivalence relation $\mathcal{U}^m$. Then we claim that

$$(y_1, y', v, w)\mathcal{U}^m(z_1, z', \bar{v}, \bar{w}) \Leftrightarrow v = \bar{v} \ \& \ w = \bar{w},$$

which clearly shows that it is tame.

$\Rightarrow$: is obvious.

$\Leftarrow$: If $v = \bar{v}, w = \bar{w}$, then $v = g_m(y_1, y') = g_m(z_1, z') = \bar{v}$, so, by (b) if $[(y_1, y')]_{E_1 \times E'} = C = C_1 \times C', [(z_1, z')]_{E_1 \times E'} = D = D_1 \times D'$, we have that $C_1 = D_1$, and since $w = \bar{w}$ and $w \in C_2, \bar{w} \in D_2$, we also have $C_2 = D_2$. So $C = D$. Thus $(y_1, y')(E_1 \times E')(z_1, z')$ and clearly

$$((y_1, y'), (z_1, z')) \cdot (z_1, z', \bar{v}, \bar{w}) = (y_1, y', \bar{v}, \bar{w}) = (y_1, y', v, w),$$

so $(y_1, y', v, w)\mathcal{U}^m(z_1, z', v, w)$. $\dashv$

Remark. Recently Gaboriau proved a result about ergodic dimension and l^2-Betti numbers that, in particular, implies that if, in the context of 7.1, $\beta_n(\Gamma_1) \neq 0$, then for any free Borel action of Γ on X with invariant measure, E_Γ^X cannot be Borel reduced to a $\mathcal{K}_n$-structurable equivalence relation. In particular, this also gives another proof of 7.1, when $\beta_n(\Gamma_1) \neq 0$, and also 7.2.

CHAPTER 8

A Final Application

Fix below an infinite locally finite countable group Z (i.e., Z is an increasing union of a sequence of finite subgroups). We will consider the groups $F_2^n \times Z, F_2^n$, for $n = 1, 2, \ldots$ and the following two free actions:

(i) The shift action of $F_2^n \times Z$ (resp., F_2^n) on $(2)^{F_2^n \times Z}$, (resp., $(2)^{F_2^n}$), whose corresponding equivalence relation is $F(F_2^n \times Z, 2)$ (resp., $F(F_2^n, 2)$).

(ii) The product of n copies of the shift action of F_2 on $(2)^{F_2}$ and the action of Z on $(2)^Z$ (resp., the product of n copies of the shift action of F_2 on $(2)^{F_2}$), whose corresponding equivalence relation is $F(F_2, 2)^n \times F(Z, 2)$ (resp., $F(F_2, 2)^n$), which up to bireducibility, $\sim_B$, is the same as $(E_{\infty T})^n \times E_0$ (resp., $(E_{\infty T})^n$), since $E(Z, 2) \sim_B E_0$; see [DJK, 7.1].

It should be pointed out that $F(F_2^n \times Z, 2)$ is the largest, in the sense of $\leq_B$, equivalence relation induced by a free Borel action of $F_2^n \times Z$, see [JKL, 3.16]. Therefore

$$(E_{\infty T})^n \times E_0 \leq_B F(F_2^n \times Z, 2).$$

Similarly

$$(E_{\infty T})^n \leq_B F(F_2^n, 2)$$

It is also clear, from [JKL, 3.16] again, that $(E_{\infty T})^n \leq_B (E_{\infty T})^n \times E_0 \leq_B (E_{\infty T})^{n+1}$, and $F(F_2^n, 2) \leq_B F(F_2^n \times Z, 2) \leq_B F(F_2^{n+1}, 2)$, for $n \geq 1$. It was also known that $E_{\infty T} <_B E_{\infty T} \times E_0$ (see [JKL, 3.28]) and $(E_{\infty T})^n < E_\infty =$ the universal countable Borel equivalence relation (see [HK, 10.8]). The next theorem is a corollary of earlier results in this paper and results of Gaboriau.

Theorem 8.1. *i)*

$$E_{\infty T} <_B E_{\infty T} \times E_0 <_B (E_{\infty T})^2 <_B \cdots <_B (E_{\infty T})^n <_B (E_{\infty T})^n \times E_0$$
$$<_B (E_{\infty T})^{n+1} <_B \cdots$$

ii)

$$(F(F_2, 2) \sim_B)\ E_{\infty T} <_B F(F_2 \times Z, 2) <_B F(F_2^2, 2) <_B \ldots$$
$$<_B F(F_2^n, 2) <_B F(F_2^n \times Z, 2) <_B F(F_2^{n+1}, 2) <_B \ldots$$

iii) Finally, for $n \geq 1$, $(E_{\infty T})^n \leq_B F(F_2^n, 2)$ and $(E_{\infty T})^n \times E_0 \leq_B F(F_2^n \times Z, 2)$, but $F(F_2 \times Z, 2) \not\leq_B (E_{\infty T})^n$, $(E_{\infty T})^n \not\leq_B F(F_2^{n-1} \times Z, 2)$, and $(E_{\infty T})^n \times E_0 \not\leq_B F(F_2^n, 2)$. In particular, $F(F_2^n, 2), (E_{\infty T})^m$ are incomparable, in $\leq_B$, if $2 \leq n < m$.

Proof. i) That $(E_{\infty T})^n <_B (E_{\infty T})^n \times E_0, n \geq 1$, follows from 7.2. Also $(E_{\infty T})^n \times E_0 <_B (E_{\infty T})^{n+1}, n \geq 1$, follows from results of Gaboriau [Ga01]. Indeed, write $E_0 = \bigcup_m R_m$, with R_m an increasing sequence of finite Borel equivalence relations on $2^{\mathbb{N}}$. Then $S_m = E_{\infty T}^n \times R_m$ is also increasing and $\bigcup_m S_m = E_{\infty T}^n \times E_0$. Now $E_{\infty T}^n \times R_m$ can be Borel reduced to a $\mathcal{K}_n$-structurable countable Borel equivalence relation (see Appendix D). If $(E_{\infty T})^{n+1} \leq_B (E_{\infty T})^n \times E_0$, towards a contradiction, say via a Borel reduction ρ, put $T_m = \rho^{-1}(S_m)$, so that T_m is increasing and $\bigcup_m T_m = (E_{\infty T})^{n+1}$. Also T_m can be Borel reduced to a $\mathcal{K}_n$-structurable countable Borel equivalence relation, so by Appendix D and [Ga01, 2.1, 2.2, 3.5] this implies, in Gaboriau's terminology, that the dimension of each T_m (with respect to the product measure on $((2)^{F_2})^{n+1}$) is $\leq n$. But then by [Ga01, 5.13], $\beta_p((E_{\infty T})^{n+1}) = 0, \forall p > n$, so, in particular, $\beta_{n+1}((E_{\infty T})^{n+1}) = 0$, thus by [Ga01, 3.16], $\beta_{n+1}(F_2^{n+1}) = \beta_n((E_{\infty T})^{n+1}) = 0$, contradicting the fact that $\beta_{n+1}(F_2^{n+1}) = 1$.

ii) $F(F_2^n, 2) <_B F(F_2^n \times Z, 2)$ follows from 3.14. We next prove that $F(F_2^n \times Z, 2) <_B F(F_2^{n+1}, 2)$. Indeed assume that $F(F_2^{n+1}, 2) \leq_B F(F_2^n \times Z, 2)$ towards a contradiction, via a Borel reduction ρ.

First, clearly $F(F_2^n \times Z, 2) = \bigcup_k R_k$, where R_k is an increasing sequence of countable Borel equivalence relations induced by a free action of a group of the form $F_2^n \times Z_k$, where Z_k is finite. Consider the Cayley graph of F_2 as a 1-dimensional contractible simplicial complex and its n-fold product (see the last paragraph of Appendix D), which is an n-dimensional contractible simplicial complex. View this as a structure $\mathcal{A}_n$ in $\mathcal{K}_n$. Clearly F_2 acts freely by automorphisms on $\mathcal{A}_n$, and thus $F_2^n \times Z_k$ acts by automorphisms on $\mathcal{A}_n$ with finite stabilizers, where we let Z_k act trivially. Then by the construction in [JKL, 3.2] R_k can be Borel reduced to a $\mathcal{K}_n$-structurable equivalence relation, and thus so can $\rho^{-1}(R_k)$. Thus $F(F_2^{n+1}, 2)$ can be written as the union of an increasing sequence of equivalence relations S_k, $F(F_2^{n+1}, 2) = \bigcup_k S_k$, where each S_k is Borel reducible to a $\mathcal{K}_n$-structurable Borel equivalence relation. This yields a contradiction, as in i) above.

iii) We have $F(F_2 \times Z, 2) \not\leq_B (E_{\infty T})^n$, by 3.14 or 6.2. The proof that $(E_{\infty T})^n \not\leq_B F(F_2^{n-1} \times Z, 2)$ is similar to the proofs in i), ii) above. Finally $(E_{\infty T})^n \times E_0 \not\leq_B F(F_2^n, 2)$ for $n \geq 1$, follows from the recent (unpublished) result of Gaboriau that the ergodic dimension (see [Ga01]) of $F_2^n \times \mathbb{Z}$ is $n+1$, which implies that $(E_{\infty T})^n \times E_0$ cannot be Borel reduced to a $\mathcal{K}_n$-structurable countable Borel equivalence relation. $\dashv$

We do not know if $F(F_2^n \times Z, 2) \sim_B F(F_2^n \times \mathbb{Z}, 2)$, for $n \geq 1$.

APPENDIX A

Strong Notions of Ergodicity

A1. Homomorphisms and relative ergodicity

Let E, F be countable Borel equivalence relations in standard Borel spaces X, Y, resp. A map $\rho : X \to Y$ such that

$$xEy \Rightarrow \rho(x)F\rho(y)$$

is called a *homomorphism* of E to F. This induces a map $\bar{\rho} : X/E \to Y/F$. If, moreover,

$$xEy \Leftrightarrow \rho(x)F\rho(y),$$

then ρ is called a *reduction* of E to F. This means that $\bar{\rho}$ is injective.

Now suppose that μ is a measure on E. We say that (E, μ) or, for simplicity, E, is *F-ergodic*, if for any Borel homomorphism $\rho : X \to Y$ of E to F there is some $y_0 \in Y$ such that $\rho(x)Fy_0$, μ-a.e. (x). Notice that "Borel homomorphism" can be replaced by "μ-measurable homomorphism" in this definition.

Recall that E is called *ergodic* if every E-invariant Borel set is either null or co-null. Thus it is easy to check that E is ergodic iff E is F-ergodic for every tame Borel equivalence relation F. We will discuss in this appendix stronger notions of ergodicity, where instead of tame F, which are the simplest equivalence relations, we look at the next more complicated class, namely the hyperfinite ones.

Finally, we call a Borel action of a countable group Γ on X *F-ergodic* if E_Γ^X is F-ergodic.

A2. E_0-ergodicity and almost invariant sets

Recall that E_0 is the equivalence relation on $2^{\mathbb{N}}$ defined by

$$xE_0y \Leftrightarrow \exists n \forall m \geq n[x(m) = y(m)].$$

This is a universal hyperfinite Borel equivalence relation, i.e., it is Borel hyperfinite, and for any Borel hyperfinite F we have $F \leq_B E_0$. In particular, this means that E is E_0-ergodic iff E is F-ergodic for every hyperfinite F.

Ergodicity of an equivalence relation E is characterized by the non-existence of non-trivial invariant sets. It turns out that E_0-ergodicity is characterized by the non-existence of non-trivial "almost invariant" sets. We proceed to explain this concept and state precisely the result.

Let Γ be a countable group and suppose that Γ acts in a Borel way on the standard Borel space X. Suppose also that μ is a Γ-*quasi-invariant* measure on X. This means that if $A \subseteq X$ is a Borel null set, so is $\gamma \cdot A, \forall \gamma \in \Gamma$, or equivalently that for every $\gamma \in \Gamma$ and every $\epsilon > 0$, there is a $\delta > 0$ such that if $\mu(A) < \delta$, then $\mu(\gamma \cdot A) < \epsilon$.

For any $\epsilon > 0, F \subseteq \Gamma$ finite, a Borel set $A \subseteq X$ is *(ϵ, F)-invariant* if $\mu(\gamma \cdot A \triangle A) < \epsilon, \forall \gamma \in F$. The action *has non-trivial almost invariant sets* if there is $\delta > 0$ so that

$\forall\epsilon, F$ as above there is Borel $A \subseteq X$ with $\delta \leq \mu(A) \leq 1-\delta$, which is (ϵ, F)-invariant. Equivalently this means that there is a sequence of Borel sets A_n with $\mu(\gamma \cdot A_n \triangle A_n) \to 0, \forall\gamma \in \Gamma$, but $\mu(A_n)(1-\mu(A_n)) \not\to 0$.

First we note that the existence of non-trivial almost invariant sets depends only on the orbit equivalence relation E_Γ^X and not on the action. Recall that if E is a countable Borel equivalence relation on the X, then $[E]$ denotes the set of all Borel automorphisms of X which leave the E-classes invariant, i.e., a Borel automorphism f is in $[E]$ if $f(x)Ex, \forall x \in X$. We now have

Lemma A2.1. *If, in the above notation,* $\mu(\gamma \cdot A_n \triangle A_n) \to 0, \forall\gamma \in \Gamma$, *then for every* $f \in [E_\Gamma^X], \mu(f(A_n)\triangle A_n) \to 0$.

Proof. Fix $\epsilon > 0$. Then, since $f \in [E_\Gamma^X]$, we can find finitely many elements $\gamma_1, \dots, \gamma_k$ of Γ and finitely many pairwise disjoint Borel sets $X_1, \dots X_k$ of X with $f(x) = \gamma_i \cdot x$ for $x \in X_i$, such that $\mu(f(X \setminus \bigcup_{i=1}^k X_i)) \leq \epsilon/4$. Choose then N large enough so that $\mu(\gamma_i \cdot A_n \triangle A_n) \leq \epsilon/4k$ and hence $\mu(f(\gamma_i^{-1} \cdot A_n \triangle A_n)) \leq \epsilon/4k$, for $i = 1, \dots, k$ and $n > N$. Fix such an $n > N$. If $x \in f(A_n) \setminus A_n$, say $x = f(y), y \in A_n$, then either $y \in X \setminus (\bigcup_{i=1}^k X_i)$, so $x \in f(X \setminus (\bigcup_{i=1}^k X_i))$, or else $y \in X_i, i = 1, \dots, k$, and so $x = f(y) = \gamma_i \cdot y$, thus $x \in \gamma_i \cdot A_n \setminus A_n$. So

$$f(A_n) \setminus A_n \subseteq \bigcup_{i=1}^k (\gamma_i \cdot A_n \setminus A_n) \cup f(X \setminus \bigcup_{i=1}^k X_i).$$

Similarly,

$$\begin{aligned} A_n \setminus f(A_n) &= f(f^{-1}(A_n) \setminus A_n) \\ &\subseteq \bigcup_{i=1}^k f(\gamma_i^{-1} \cdot A_n \setminus A_n) \cup f(X \setminus \bigcup_{i=1}^k X_i). \end{aligned}$$

So,

$$\begin{aligned} \mu(A_n \triangle f(A_n)) \leq \sum_{i=1}^k \mu(\gamma_i \cdot A_n \setminus A_n) + \sum_{i=1}^k \mu(f(\gamma_i^{-1} \cdot A_n \setminus A_n)) \\ + 2\mu(f(X \setminus \bigcup_{i=1}^k X_i)) \leq \epsilon. \end{aligned}$$

⊣

It follows that if E is a countable Borel equivalence relation on a standard Borel space X and μ an E-quasi-invariant measure (i.e., the E-saturation of a null set in null), then we can define unambiguously the notion: *E has non-trivial almost invariant sets.* It simply means that some Borel action of a countable group Γ that induces E has non-trivial almost invariant sets, and this is equivalent to saying that *every* Borel action of a countable group Γ that induces E has non-trivial almost invariant sets, and also equivalent to the statement that there is a sequence A_n of Borel sets with $\mu(f(A_n)\triangle A_n) \to 0, \forall f \in [E]$, but with $\mu(A_n)(1-\mu(A_n) \not\to 0$.

It is also easy to see that this notion does not depend on the measure μ but only on the measure class of μ.

We now have the following result:

Theorem A2.2 (Jones-Schmidt [JS]). *Let E be a countable Borel equivalence relation on a standard Borel space X and let μ be a measure on X. Assume that E is ergodic and μ is E-quasi-invariant. Then the following are equivalent:*

(i) E is E_0-ergodic.

(ii) E fails to have non-trivial almost invariant sets.

Proof. Fix a countable group Γ and a Borel action of Γ on X with $E = E_\Gamma^X$.

(i) $\Rightarrow$ (ii): Assume (ii) fails. Then fix $\delta > 0$ and a sequence A_n of Borel sets with $\delta \leq \mu(A_n) \leq 1 - \delta$, such that for any given $\gamma \in \Gamma$, we have $\mu(\gamma \cdot A_n \triangle A_n) \leq 2^{-n}$, for all sufficiently large n. We will find a Borel homomorphism ρ of E to E_0 with $\rho^{-1}([y]_{E_0})$ null for each $y \in Y$. Then ρ shows that E is not E_0-ergodic.

For each fixed $\gamma \in \Gamma$, and all large enough n,

$$A_n^\gamma = \gamma^{-1} \cdot A_n \triangle A_n$$

has measure $\leq 2^{-n}$, thus the set

$$A^\gamma = \bigcap_m \bigcup_{m \leq n} A_n^\gamma$$

has measure 0, so for each $\gamma \in \Gamma$ and almost all $x, x \in \bigcup_m \bigcap_{n \geq m} (X \setminus A_n^\gamma)$, i.e., for almost all x,

$$(*) \qquad \exists m \forall n \geq m (x \in A_n \Leftrightarrow \gamma \cdot x \in A_n).$$

So fix an E-invariant co-null set $X_0 \subseteq X$ such that for each $x \in X_0, \gamma \in \Gamma$, we have that $(*)$ holds.

Now define $\rho : X \to 2^{\mathbb{N}}$ by

$$\rho(x) = \begin{cases} n \mapsto 1_{A_n}(x), & \text{if } x \in X_0, \\ a, & \text{if } x \notin X_0, \end{cases}$$

where a is some fixed element of $2^{\mathbb{N}}$ and 1_A = characteristic function of A. From $(*)$ it is immediate that ρ is a Borel homomorphism of E to E_0. We finally check that $\rho^{-1}([y]_{E_0})$ is null, for each $y \in 2^{\mathbb{N}}$.

We have

$$B = \rho^{-1}([y]_{E_0}) = \bigcup_m \bigcap_{n \geq m} A_n^{y(n)},$$

where $A_n^1 = A_n, A_n^0 = X \setminus A_n$. Since B is E-invariant, and E is ergodic, it is enough to check that $\mu(B) < 1$. But clearly $\mu(\bigcap_{n \geq m} A_n^{y(n)}) \leq 1 - \delta$, since $\mu(A_n), \mu(X \setminus A_n) \leq 1 - \delta$, so $\mu(B) \leq 1 - \delta$, and the proof is complete.

(ii) $\Rightarrow$ (i): Assume (i) fails and fix a Borel homomorphism $\pi : X \to 2^{\mathbb{N}}$ from E to E_0 which does not map a co-null set into a single E_0-class. We will show that E has non-trivial almost invariant sets.

Write $E_0 = \bigcup_{n=1}^\infty E_n$, where E_n are increasing *finite* Borel equivalence relations. Fix $\epsilon > 0, F \subseteq \Gamma$ symmetric (i.e., closed under inverses) finite, in order to find an (ϵ, F)-invariant set. Since the E_n are increasing, we have for each $x \in X$ some $n_x \geq 1$ with $\pi(F \cdot x) \subseteq [\pi(x)]_{E_{n_x}}$. So $X = \bigcup_{n=1}^\infty X_n$, where $X_n = \{x \in X : \pi(F \cdot x) \subseteq [\pi(x)]_{E_n}\}$. Clearly $X_1 \subseteq X_2 \subseteq \dots$, so there is some large enough n with $\mu(X \setminus X_n) < \epsilon/2$ and $\mu(\gamma \cdot (X \setminus X_n)) < \epsilon/2$ for all $\gamma \in F$. Put $Y = X_n$. Then $\mu(X \setminus Y) < \epsilon/2$ and for $x \in Y, \pi(F \cdot x) \subseteq [\pi(x)]_{E_n}$. Fix a Borel transversal, T, for

E_n. Let $\nu = \pi_*\mu$, so that $\nu(C) = 0$ for each E_n-class C. Define the measure τ on T by

$$\tau(S) = \nu([S]_{E_n}),$$

for any Borel set $S \subseteq T$. Clearly τ is non-atomic. So there is a Borel set $T_0 \subseteq T$ with $\tau(T_0) = 1/2$. Put $A = \pi^{-1}([T_0]_{E_n})$, so that $\mu(A) = 1/2$.

If $\gamma \in F, x \in A \setminus \gamma \cdot A$ and $x \in Y$, then, as $\gamma^{-1} \in F, \pi(\gamma^{-1} \cdot x) \in [\pi(x)]_{E_n} \subseteq [T_0]_{E_n}$, so $\gamma^{-1} \cdot x \in A$, or $x \in \gamma \cdot A$, a contradiction. So $A \setminus \gamma \cdot A \subseteq X \setminus Y$. Similarly, $\gamma \cdot A \setminus A = \gamma \cdot (A \setminus \gamma^{-1} \cdot A) \subseteq \gamma \cdot (X \setminus Y)$. So $\gamma \cdot A \triangle A \subseteq (X \setminus Y) \cup \gamma \cdot (X \setminus Y)$ and thus $\mu(\gamma \cdot A \triangle A) < \epsilon, \forall \gamma \in F$, i.e., A is (ϵ, F)-invariant. ⊣

Remark A2.3. Notice that the proof of (ii)⇒(i) also shows that the sets witnessing almost invariance can be chosen to have measure c for any fixed $0 < c < 1$.

Remark A2.4. In [JS, 2.1], it is actually shown that if (ii) fails, then one can find $\rho : X \to 2^{\mathbb{N}}$, a Borel homomorphism of E to E_0, so that moreover $\rho_*\mu \sim \mu_0$, where μ_0 is the usual product measure on $2^{\mathbb{N}}$, and if μ is actually E-invariant we can insure that $\rho_*\mu = \mu_0$.

A3. Almost invariant vectors

Consider now a Borel action of a countable group Γ on a standard Borel space X with invariant measure μ. An *I-sequence* is a sequence A_n of Borel sets with $\mu(A_n) > 0$, $\mu(A_n) \to 0$, such that $\frac{\mu(A_n \triangle \gamma \cdot A_n)}{\mu(A_n)} \to 0, \forall \gamma \in G$ (see [dJR], [R], [S81]). If this action has non-trivial almost invariant sets, then, as we discussed in A2.3, for each $0 < c < 1$ there is a sequence B_n^c with $\mu(\gamma \cdot B_n^c \triangle B_n^c) \to 0, \forall \gamma \in G$ and $\mu(B_n^c) = c$. Let then, for each $n, A_n = B_{k_n}^{1/n}$, where k_n is large enough so that $\mu(\gamma_i \cdot B_{k_n}^{1/n} \triangle B_{k_n}^{1/n}) < \frac{1}{n^2}$, for $i = 1, \dots, n$, where $\Gamma = \{\gamma_i\}_{i=1}^{\infty}$. Clearly A_n is an I-sequence. So we have:

Proposition A3.1. *If the action has non-trivial almost invariant sets, then it has an I-sequence.*

Schmidt [S81, 2.7] has a counterexample to show that the converse fails, for some ergodic measure preserving action of F_3. In fact one can prove the following stronger result.

Theorem A3.2. *Let the free group $F_2 = \langle a, b \rangle$ act in a Borel way freely on a standard Borel space X with invariant measure μ. Let $T_a(x) = a \cdot x$ be the Borel automorphism of X corresponding to the generator a of F_2. Assume that μ is ergodic with respect to T_a. Then there is another free action of F_2 on X, which gives the same orbit equivalence relation on an F_2-invariant Borel set of μ measure 1, and such that this new action admits an I-sequence.*

The hypotheses of this theorem are satisfied if the given action of F_2 is mixing, like, e.g., the free part of the shift action of F_2 on 2^{F_2} with the usual product measure μ (see A6.1). This shift action does not have an I-sequence (so in particular it does not have non-trivial almost invariant sets). This follows from A4.1 and A3.4. However, by A3.2, one can modify it to another action of F_2, on the same space, generating the same equivalence relation, μ-a.e., so that this new action has an I-sequence but no non-trivial almost invariant sets. In particular, this shows that the existence of an I-sequence is a property of the action and not the equivalence relation.

Proof of A3.2. We will use the following:

Lemma A3.3. *There is a standard Borel space Y, a measure ν on Y and a measure preserving, ergodic, aperiodic (i.e., having no finite orbits) Borel automorphism S of Y, with the following property. For each $n = 1, 2, \dots$ there is a Borel set A_n of positive measure such that*

(1) $\frac{\nu(A_n \triangle S(A_n))}{\nu(A_n)} \leq \frac{1}{n}$, $\nu(A_n) \leq 2^{-n}$,

(2) $\forall |i| \leq n \forall |j| \leq m (S^i(A_n) \cap S^j(A_m) = \emptyset)$, *if* $n \neq m$.

Assuming this, we complete the proof as follows:

The equivalence relations induced by T_a, S are Borel isomorphic, when restricted to invariant sets of measure 1, by an isomorphism that sends μ to ν, by Dye's Theorem (see [D], [Zi84, 4.3.12]). Neglecting null sets, denote by S_a the Borel automorphism of X corresponding to S by this isomorphism, and B_n the sets in X corresponding to A_n. Thus, modulo null sets, S_a generates the same equivalence relation as T_a, and $\frac{\mu(B_n \triangle S_a(B_n))}{\mu(B_n)} \leq \frac{1}{n}$, $\mu(B_n) \leq 2^{-n}$, and $\forall |i| \leq n \forall |j| \leq m(S_a^i(B_n) \cap S_a^j(B_m) = \emptyset)$, if $n \neq m$.

Now consider any atom C of the Boolean algebra on $X_n = \bigcup_{|i| \leq n} S_a^i(B_n)$ generated by the sets $\{S_a^i(B_n) : |i| \leq n\}$, such that C has positive measure. Since $\mu(b \cdot C) = \mu(C)$, and μ is S_a-ergodic, there is φ_C, a Borel automorphism of X, with $\varphi_C(x) = S_a^{n(x)}(x)$, for some $n(x) \in \mathbb{Z}$, μ-a.e., and $\varphi_C(b \cdot C) = C$. Let $\psi_C : C \to C$ be defined by $\psi_C(x) = \varphi_C(b \cdot x)$.

Similarly if $D = X \setminus \bigcup_n X_n$ define φ_D, ψ_D. Finally, let $S_b = \psi_D \cup \bigcup_n \{\psi_C : C$ is an atom of X_n of positive measure$\}$. Clearly S_b is a Borel automorphism of X, μ-a.e., and $S_b(S_a^i(B_n)) = S_a^i(B_n)$ for each n and $|i| \leq n$. Consider the action of F_2 on X, where a acts by S_a and b by S_b. Clearly this action generates the same equivalence relation as the original one, μ-a.e. Moreover, it is free a.e. by Gaboriau's [Ga00] results on costs: The equivalence relation has cost 2, and the new action gives a graphing of cost at most 2, thus exactly 2, so it has to be a treeing, i.e., it is free a.e.

Finally, we check that $\{B_n\}$ is an I-sequence for the new action. Fix $\gamma \in F_2$. Say $\gamma = a^{i_1} b^{j_1} \dots a^{i_k} b^{j_k}$. Fix $n > |i_1| + |i_2| + \dots + |i_k|$. Then $S_b^{j_k}(B_n) = B_n$,

$$S_a^{i_k}(S_b^{j_k}(B_n)) = S_a^{i_k}(B_n),$$

$$S_b^{j_{k-1}}(S_a^{i_k} S_b^{j_k}(B_n)) = S_b^{j_{k-1}}(S_a^{i_k}(B_n)) = S_a^{i_k}(B_n),$$

as $|i_k| < n$, etc., and so $\gamma \cdot B_n = S_a^{i_1 + i_2 + \dots + i_k}(B_n)$. So

$$\mu(\gamma \cdot B_n \triangle B_n) = \mu(S_a^{i_1 + \dots + i_k}(B_n) \triangle B_n) \leq (|i_1| + \dots + |i_k|) \mu(S_a(B_n) \triangle B_n).$$

Thus clearly

$$\frac{\mu(\gamma \cdot B_n \triangle B_n)}{\mu(B_n)} \to 0,$$

as $n \to \infty$.

So it is enough to give the

Proof of Lemma A3.3. Consider the odometer map $\mathcal{W} : 2^{\mathbb{N}} \to 2^{\mathbb{N}}$ given by $\mathcal{W}(1^n 0 x) = 0^n 1 x$, and the usual product measure ρ on $2^{\mathbb{N}}$. $\mathcal{W}$ is measure preserving and ergodic. Let $C^n = \{1^n 0 x : x \in 2^{\mathbb{N}}\}$, so that $\mathcal{W}(C^n) = \{0^n 1 x : x \in 2^{\mathbb{N}}\}$. Clearly $\rho(C^n) = \frac{1}{2^{n+1}}$, and if $1 \leq m < n$, then the family $\{C^m, \mathcal{W}(C^m), C^n, \mathcal{W}(C^n)\}$ is pairwise disjoint.

For each n, let $X_0^n = C^{2n}, X_1^n, \ldots X_{2^{n+1}}^n, X_{2^{n+1}+1}^n = \mathcal{W}(C^{2n})$ be disjoint copies of C^{2n} and fix Borel bijections $\mathcal{W}_i^n : X_i^n \to X_{i+1}^n, 0 \leq i \leq 2^{n+1}$. Let Y be the disjoint union of $2^{\mathbb{N}}$ and the sets $X_1^n, X_2^n, \ldots X_{2^{n+1}}^n, n = 1, 2, \ldots$ Define S on Y as follows: $S|(2^{\mathbb{N}} \setminus \bigcup_{n=1}^{\infty} C^{2n}) = \mathcal{W}|(2^{\mathbb{N}} \setminus \bigcup_{n=1}^{\infty} C^{2n}), S|X_i^n = \mathcal{W}_i^n, 0 \leq i \leq 2^{n+1}, n = 1, 2, \ldots$ Define the measure $\bar{\nu}$ on Y, so that $\bar{\nu}|2^{\mathbb{N}} = \rho$ and $\bar{\nu}$ on X_i^n is a copy of $\rho|C^{2n}$ in the obvious way, so that each $\mathcal{W}_i^n$ is measure preserving. Of course, $\bar{\nu}$ is not a probability measure, but $\bar{\nu}(Y) = 1 + \sum_{n=1}^{\infty} \frac{2^{n+1}}{2^{2n+1}} = 2$, so put $\nu = \frac{\bar{\nu}}{2}$. It is easy to see that S is a measure preserving Borel automorphism on (Y, ν) and it is ergodic (since $X \subseteq Y$ is a complete section for the equivalence relation given by S, and this equivalence relation restricted to X is the one given by $\mathcal{W}$). Finally, take for $n \geq 3$,

$$A'_n = \bigcup_{i=-n}^{n} X_{2^n+i}^n,$$

and let $A_n = A'_{n+2}$. This clearly works. ⊣

If a countable group Γ acts in a Borel way on a standard Borel space X with invariant measure μ, then Γ acts unitarily on $L^2(X, \mu)$ by

$$\gamma \cdot f(x) = f(\gamma^{-1} \cdot x).$$

Let $L_0^2(X, \mu)$ be the closed subspace of $L^2(X, \mu)$ defined by

$$L_0^2(X, \mu) = \{f \in L^2(X, \mu) : \int f d\mu = 0\}.$$

This is clearly the orthogonal complement of the closed subspace of $L^2(X, \mu)$ consisting of the constant functions. Of course $L_0^2(X, \mu)$ is also Γ-invariant. Finally note that μ is Γ-ergodic iff the only invariant vectors for the action of Γ on $L^2(X, \mu)$ are the constant functions iff the action of Γ on $L_0^2(X, \mu)$ has no non-0 invariant vectors.

In general, if a countable group Γ acts unitarily on a Hilbert space H, and $\epsilon > 0, F \subseteq \Gamma$ is finite, then a non-0 vector $x \in H$ is called *(ϵ, F)-invariant* if $\frac{||\gamma \cdot x - x||}{||x||} < \epsilon, \forall \gamma \in F$. The action has *non-0 almost invariant vectors* if $\forall \epsilon, F$ as above there is a non-0 (ϵ, F)-invariant vector (depending of course on (ϵ, F)). This is, as usual, equivalent to the existence of a sequence x_n of non-0 vectors in H with $\frac{||\gamma \cdot x_n - x_n||}{||x_n||} \to 0, \forall \gamma \in \Gamma$.

Recall also that in the context of a Borel action of Γ on (X, μ) with μ Γ-invariant, a *Γ-invariant mean* on $L^\infty(X, \mu)$ is a linear functional $M : L^\infty(X, \mu) \to \mathbb{C}$ with $M(1) = 1, M(f) \geq 0, \forall f \geq 0$, and $M(\gamma \cdot f) = M(f), \forall f \in L^\infty(X, \mu), \gamma \in \Gamma$ (as usual Γ acts on $L^\infty(X, \mu)$ by $\gamma \cdot f(x) = f(\gamma^{-1} \cdot x)$). Clearly integration by μ, i.e., $M(f) = \int f d\mu$ is a Γ-invariant mean on $L^\infty(X, \mu)$.

It turns out that the existence of an I-sequence is equivalent to other important properties of the action of Γ on (X, μ) and the corresponding unitary action of Γ on $L_0^2(X, \mu)$.

Theorem A3.4 (Rosenblatt [R], Schmidt [S81]). *Let Γ be a countable group acting in a Borel way on a standard Borel space X with invariant, ergodic measure μ. Then the following are equivalent:*

(i) The action has an I-sequence.

(ii) The action of Γ on $L_0^2(X, \mu)$ has non-0 almost invariant vectors.

(iii) There is more than one Γ-invariant mean on $L^\infty(X, \mu)$.

The equivalence of (i), (iii) is proved in [R, 1.4]. The equivalence of (i), (ii) is proved in [S81, 2.3]. (The implication (i)⇒(ii) is quite easy: If A_n is an I-sequence, let $f_n = 1_{A_n} - \mu(A_n)$, so that $f_n \in L^2_0(X,\mu)$. Then $\frac{||\gamma \cdot f_n - f_n||}{||f_n||} \to 0, \forall \gamma \in \Gamma$, so the unitary action of Γ on $L^2_0(X,\mu)$ has non-0 almost invariant vectors.)

In particular, of course, if the action of Γ on X has non-trivial almost invariant sets, by A3.1 it has an I-sequence, so the action of Γ on $L^2_0(X,\mu)$ has non-0 almost invariant vectors. This last conclusion is however easy to prove directly from the existence of non-trivial almost invariant sets. Indeed, fix $\delta > 0$ and for each ϵ, F choose Borel $A \subseteq X$ with $\delta < \mu(A) < 1 - \delta$ which is $(\epsilon^2\delta^2, F)$-invariant. Put $f = 1_A - \mu(A)$. Then $f \in L^2_0(X,\mu), ||f|| > \delta$ and $||\gamma \cdot f - f|| = \sqrt{\mu(\gamma \cdot A \Delta A)} < \epsilon\delta$, so $\frac{||\gamma \cdot f - f||}{||f||} < \epsilon$, i.e., f is (ϵ, F)-invariant.

To conclude this section, we recall how the notion of almost invariant vectors can be used to characterize amenability.

For each countable group Γ, the *regular action* (or *representation* of Γ) is the action of Γ on $\ell^2(\Gamma)$ given by

$$\gamma \cdot x(\delta) = x(\gamma^{-1}\delta).$$

Theorem A3.5 (Hulanicki; see [Zi84, 7.1.8]). *A countable group Γ is amenable iff the regular action of Γ on $\ell^2(\Gamma)$ has non-trivial almost invariant vectors.*

A4. E_0-ergodicity of the shift action

We will see here that the shift action of Γ on 2^Γ is E_0-ergodic, when Γ is not amenable. (Clearly the converse is also true, since if Γ is amenable the equivalence relation $E_\Gamma^{2^\Gamma}$ is hyperfinite a.e. by [OW]). We will actually state this result in a somewhat stronger form.

Theorem A4.1 (see Losert-Rindler [LR], Jones-Schmidt [JS]). *Let Δ be a countable group and $\Gamma \leq \Delta$ a non-amenable subgroup. Consider the shift action of Γ on 2^Δ (i.e., the restriction to Γ of the shift action of Δ on 2^Δ) and equip 2^Δ with the usual product measure μ. Then the action of Γ on $L^2_0(2^\Delta,\mu)$ does not admit non-0 almost invariant vectors and thus the shift action of Γ on 2^Δ is E_0-ergodic.*

Proof. We will relate the action of Γ on $L^2_0(2^\Delta,\mu)$ to the regular action of Γ and show that if the first action had non-0 almost invariant vectors, so would the second, contradicting A3.5.

We will identify 2^Δ with $\mathbb{Z}_2^\Delta$ ($\mathbb{Z}_2 = \mathbb{Z}/2\mathbb{Z}$), which is a compact abelian group. Thus μ is simply the Haar measure of $\mathbb{Z}_2^\Delta$. The dual group of $\mathbb{Z}_2^\Delta$, i.e., its group of characters, is $\widehat{\mathbb{Z}_2^\Delta} \cong \mathbb{Z}_2^{\boldsymbol{\Delta}} = \{\chi \in \mathbb{Z}_2^\Delta : \chi(\delta) = 0$ for all but finitely many $\delta \in \Delta\}$ and the character associated to a given $\chi \in \mathbb{Z}_2^{\boldsymbol{\Delta}}$ is defined by

$$\hat{\chi}(x) = (-1)^{\sum_\delta \chi(\delta)x(\delta)},$$

for $x \in \mathbb{Z}_2^{\boldsymbol{\Delta}}$. We will not distinguish between $\chi \in \mathbb{Z}_2^{\boldsymbol{\Delta}}$ and the associated character $\hat{\chi}$ (a homomorphism from $\mathbb{Z}_2^\Delta$ into $\mathbb{T}$). Moreover $\int \chi(x)d\mu(x) = 0$ for any $\chi \in \mathbb{Z}_2^{\boldsymbol{\Delta}}, \chi \neq 0$ and $\{\chi \in \mathbb{Z}_2^{\boldsymbol{\Delta}} : \chi \neq 0\}$ form an orthonormal basis for the Hilbert space $L^2_0(\mathbb{Z}_2^\Delta,\mu)$ (see, e.g., [Ka, p. 193]). (In fact, one can see this directly for this specific case, as follows: Easy calculation show that they are an orthonormal set. To see that they span, we note that the characters separate parts and are closed under multiplication, hence their linear combinations are dense by Stone-Weierstrass.)

Consider now the action of Γ on $\{\chi \in \mathbb{Z}_2^{\boldsymbol{\Delta}} : \chi \neq 0\}$. Since every $\chi \in \mathbb{Z}_2^{\boldsymbol{\Delta}}, \chi \neq 0$, is simply a characteristic function of a finite non-$\emptyset$ subset of Δ, this is just the left-translation action of Γ on the set of finite non-$\emptyset$ subsets of Δ, thus this action has finite stabilizers. (In case Γ is torsion-free, the action is actually free.) Let $\mathcal{W}_1, \mathcal{W}_2 \dots$ be the orbits of the action of Γ on $\{\chi \in \mathbb{Z}_2^{\boldsymbol{\Delta}} : \chi \neq 0\}$ and choose $\chi_n \in \mathcal{W}_n$. Let H_n be the closed subspace of $L_0^2(X, \mu)$ with basis consisting of all $\chi \in \mathcal{W}_n$ (i.e., all $\gamma \cdot \chi_n, \gamma \in \Gamma$). Then clearly $L_0^2(\mathbb{Z}_2^\Delta, \mu) = H_1 \oplus H_2 \oplus \dots$, and each H_n is Γ-invariant. Denote by Γ_n the stabilizer of χ_n (i.e., the set $\{\gamma \in \Gamma : \gamma \cdot \chi_n = \chi_n\}$). Then Γ_n is a finite subgroup of Γ. Consider the left-coset space Γ/Γ_n, on which Γ acts by $\gamma \cdot h\Gamma_n = \gamma h \Gamma_n$. This gives a unitary action of Γ on $\ell^2(\Gamma/\Gamma_n)$ given by $\gamma \cdot f(h\Gamma_n) = f(\gamma^{-1} h \Gamma_n)$. Clearly this action is isomorphic to the action of Γ on H_n. (Two unitary actions of Γ on Hilbert spaces $\mathcal{H}_1, \mathcal{H}_2$ are *isomorphic* if there is a Hilbert space isomorphism $\varphi : \mathcal{H}_1 \to \mathcal{H}_2$ such that $\varphi(\gamma \cdot x) = \gamma \cdot \varphi(x), \forall x \in \mathcal{H}_1$.)

Now assume, towards a contradiction, that the action of Γ on $L_0^2(\mathbb{Z}_2^\Delta, \mu)$ has non-0 almost invariant vectors. We will show that for each (ϵ, F) there is an $n \geq 1$ such that the action of Γ on H_n, and thus on $\ell^2(\Gamma/\Gamma_n)$ has an (ϵ, F)-invariant vector. We will then easily deduce from this that $\ell^2(\Gamma)$ has an (ϵ, F)-invariant vector, i.e., $\ell^2(\Gamma)$ has non-0 almost invariant vectors, a contradiction.

So fix $\epsilon > 0, F \subseteq \Gamma$ finite. Find then $x \neq 0$ in $L_0^2(\mathbb{Z}_2^\Delta, \mu)$ with $\frac{||\gamma \cdot x - x||}{||x||} < \epsilon/F, \forall \gamma \in F$. Let $x = x_1 + x_2 + \dots$, with $x_n \in H_n$. Then $\sum_{\gamma \in F} ||\gamma \cdot x - x||^2 < \epsilon^2 \cdot ||x||^2$, so

$$\sum_{\gamma \in F} \sum_n ||\gamma \cdot x_n - x_n||^2 < \epsilon^2 \cdot \sum_n ||x_n||^2$$

or

$$\sum_n \sum_{\gamma \in F} ||\gamma \cdot x_n - x_n||^2 < \epsilon^2 \sum_n ||x_n||^2,$$

so, for some n,

$$\sum_{\gamma \in F} ||\gamma \cdot x_n - x_n||^2 < \epsilon^2 ||x_n||^2,$$

thus

$$\frac{||\gamma \cdot x_n - x_n||^2}{||x_n||^2} < \epsilon^2, \text{ for all } \gamma \in F,$$

i.e., x_n is (ϵ, F)-invariant.

So we have seen that $\ell^2(\Gamma/\Gamma_n)$ has an (ϵ, F)-invariant vector, for which we can of course assume that it has norm 1. Call it v. We will find an (ϵ, F)-invariant vector for $\ell^2(\Gamma)$ as follows:

Let $u \in \ell^2(\Gamma)$ be defined by

$$u(h) = \frac{1}{\sqrt{|\Gamma_n|}} v(h\Gamma_n).$$

Then $||u|| = 1$ and $||\gamma \cdot u - u|| = ||\gamma \cdot v - v|| < \epsilon, \forall \gamma \in F$, so we are done. ⊣

A5. Characterizations of amenable and Kazhdan groups

Theorem A5.1 (Schmidt [S81]). *Let Γ be a countable group. Then the following are equivalent:*

(i) Γ is amenable.

(ii) Every Borel action of Γ on a standard Borel space X with quasi-invariant, ergodic measure μ has non-trivial almost invariant sets.

(iii) Every Borel action of Γ *on a standard Borel space* X *with invariant, ergodic measure* μ *has an* I*-sequence.*

(i) $\Rightarrow$(ii) follows from the Ornstein-Weiss Theorem [OW], which implies that E_Γ^X is hyperfinite, μ-a.e., and A2.2. (ii)$\Rightarrow$(iii) is clear from A3.1. Finally, $\neg$(i)$\Rightarrow$ $\neg$(iii) follows from A4.1 and A3.4, (i)$\Rightarrow$(ii).

Next we have a characterization of Kazhdan groups.

Theorem A5.2 (Connes-Weiss [CW], Schmidt [S81]). *Let* Γ *be a countable group. Then the following are equivalent:*

(i) Γ *is Kazhdan.*

(ii) Every Borel action of Γ *on a standard Borel space* X *with invariant, ergodic measure* μ *does not have non-trivial almost invariant sets (i.e., it is* E_0*-ergodic).*

(iii) Every Borel action of Γ *on a standard Borel space* X *with invariant, ergodic measure* μ *does not have an* I*-sequence (i.e., the corresponding action of* Γ *on* $L_0^2(X,\mu)$ *does not have non-0 almost invariant vectors).*

Recall that Γ is a *Kazhdan group* iff any unitary action of Γ on a separable Hilbert space H, which has non-0 almost invariant vectors, actually has non-0 invariant vectors. Thus (i)$\Rightarrow$(iii) is immediate. $\neg$(ii)$\Rightarrow$ $\neg$(iii) follows from A3.1. Finally (ii)$\Rightarrow$(i) is proved in Connes-Weiss [CW].

A6. Mixing

Another strong ergodicity property that is considered in ergodic theory is the concept of mixing.

Let Γ be any infinite countable group acting in a Borel way on a standard Borel space X with invariant measure μ. This action is *(strongly) mixing* if for any two Borel sets $A, B \subseteq X$.

$$\lim_{\gamma\to\infty} \mu(\gamma \cdot A \cap B) = \mu(A)\mu(B)$$

(here $\lim_{\gamma\to\infty} f(\gamma) = a$ means that $\forall\epsilon\exists F \subseteq \Gamma$ (F finite & $|f(\gamma) - a| < \epsilon, \forall\gamma \notin F$).

For general information about mixing, see Schmidt [S84], Schmidt-Walters [SW], and Bekka-Mayer [BM, Ch. I]).

It is clear that mixing is inherited to infinite subgroups, i.e., if the action of Γ on X is mixing so is the restriction of the action to any infinite $\Delta \leq \Gamma$. Since it is clear that any mixing action is ergodic (take $A = B$ to be an invariant set in the definition), it follows that if the action of Γ on X is mixing, the action of any infinite $\Delta \leq \Gamma$ is, in particular, ergodic. It is not however true that, conversely, if every infinite subgroup $\Delta \leq \Gamma$ acts ergodically on X, then the action of Γ is mixing. To see an example, notice first that if G is an infinite compact Polish group with its Haar measure μ and $\Gamma \leq G$ a countable dense subgroup, then the action of Γ on G by left-translation is ergodic (see e.g., [Zi84, 2.2.13]). On the other hand, this action is not mixing. Indeed, fix $\gamma_n \in \Gamma$, $\gamma_n \to g \neq 1$. Then find a compact nbhd N of $1 \in G$ such that $N \cap gN = \emptyset$. Then for large enough n, $N \cap \gamma_n N = \emptyset$. So, if this was mixing, $\mu(N)^2 = \mu(N)\mu(\gamma_n N) = \lim_{n\to\infty} \mu(N \cap \gamma_n N) = 0$, a contradiction. In particular, if $G = \mathbb{T}$, Γ is a dense subgroup of $\mathbb{T}$ isomorphic to $\mathbb{Z}$, then every infinite subgroup of Γ is also dense, thus the translation action of Γ on $\mathbb{T}$ has the property that every infinite subgroup acts ergodically but the action is not mixing.

The primary example of a mixing action of Γ is its shift action on 2^Γ. The following is well-known.

Proposition A6.1. *The shift action of an infinite countable group* Γ *on* 2^Γ, *with the usual product measure, is mixing.*

Proof. The basic clopen sets in 2^Γ are the sets

$$N_s = \{f \in 2^\Gamma : f|\text{dom}(s) = s\},$$

where s varies over all maps $s : F \to \{0,1\}$, with F a finite subset of Γ. Since finite unions of basic clopen sets are dense in the measure algebra of μ, it is enough to verify the mixing condition for A, B finite unions of basic clopen sets. If N_s is as above, we call F the *support* of N_s. If $A = N_{s_1} \cup \cdots \cup N_{s_k}$ and F_i is the support of N_{s_i}, we call $F_1 \cup \cdots \cup F_k$ a support of A. Fix then F_A, F_B supports for A, B resp. Then, except for finitely many $\gamma, \gamma F_A \cap F_B = \emptyset$, so $\gamma \cdot A, B$ are independent sets, thus

$$\mu(\gamma \cdot A \cap B) = \mu(\gamma \cdot A)\mu(B) = \mu(A)\mu(B).$$

$\dashv$

It is easy to verify that mixing is preserved under diagonal actions: If the actions of Γ on $(X_i, \mu_i), i = 1, 2, \ldots$ are mixing, so is the diagonal action

$$\gamma \cdot (x_i) = (\gamma \cdot x_i)$$

on $(\prod_i X_i, \prod_i \mu_i)$. Another important property is that if an action of Γ on (X, μ) is mixing and an action of Γ on (Y, ν) is ergodic, then the diagonal action of Γ on $(X \times Y, \mu \times \nu)$ is ergodic, see [SW, 2.3] or [S84].

Finally, we point out that the two strong notions of ergodicity that we have discussed in this appendix are not related. There are E_0-ergodic actions that are not mixing and there are mixing actions that are not E_0-ergodic.

For the first type, consider the standard action (by matrix multiplication) of $\text{SL}_2(\mathbb{Z})$ on $\mathbb{T}^2$ (equipped with the product measure). In [S80, 3.6] it is shown that this action admits no non-trivial almost invariant sets, so it is E_0-ergodic. (This can be also proved as follows: It is enough to show that the action of $SL_2(\mathbb{Z})$ on $L_0^2(\mathbb{T}^2)$ admits no non-0 almost invariant vectors. To see this, find a copy of F_2 contained in $SL_2(\mathbb{Z})$ so that the non-identity matrices in this copy do not have 1 as an eigenvalue (see [Wa, p. 86]), thus act freely by matrix multiplication on $\mathbb{Z}^2 \setminus \{(0,0)\}$. As in the proof of A4.1 (since a basis for $L_0^2(\mathbb{T}^2)$ is given by the non-trivial characters, i.e., the elements of $\mathbb{Z}^2 \setminus \{(0,0)\}$, on which F_2 acts freely), it follows that the restriction to this copy of F_2 of the action of $SL_2(\mathbb{Z})$ on $L_0^2(\mathbb{T}^2)$ gives a representation, which can be decomposed as a direct sum of subrepresentations isomorphic to the left regular representation. As in the proof of 4.1, this shows that F_2 is amenable, a contradiction.) But it is well-known that this action is not mixing (see, e.g., [BM, 2.11 (iv)]).

One can also give a completely "soft" proof that E_0-ergodicity does not imply mixing. We begin with any measure preserving free action of F_2 on a standard Borel probability space (X, μ) which is E_0-ergodic. We let $T_a, T_b : X \to X$ be the measure preserving transformations induced by the generators of the free group. Following Dye's theorem we can find a new automorphism $S : X \to X$ of the space which has the same orbit equivalence relation as T_a but which is not mixing. Then if we go back and define a new action of F_2 with

$$a \cdot x = S(a),$$

$$b \cdot x = T_b(x),$$

then the equivalence relation remains E_0-ergodic, since it is in fact unchanged, but the action is not mixing simply because the induced action of the subgroup $\langle a\rangle \leq F_2$ is not mixing. It is also not hard to see that the action will still be free a.e.

For the second type, we can quote the following characterization of the so-called groups with the *Haagerup approximation property* (HAP), for which we refer the reader to [CCJJV] (see also [HK] where these groups were called *strongly non-Kazhdan* (SNK) groups).

Theorem A6.2 (Jolissaint, see [CCJJV, 2.1.3]). *Let Γ be a countable group. Then Γ has the HAP iff Γ admits a free Borel action on a standard Borel space X with invariant measure μ which is mixing but not E_0-ergodic.*

Since, e.g., F_2 has the HAP, this shows that F_2 has free Borel actions with invariant measure which are mixing but not E_0-ergodic.

Such an example can be also constructed by elementary means:

We will consider the shift action of F_2 on 2^{F_2} and define a sequence of invariant measures μ_n on 2^{F_2} such that the action of F_2 on $(X_n, \mu_n) = (2^{F_2}, \mu_n)$ is mixing, therefore the diagonal action of F_2 on $(\prod_n X_n, \prod_n \mu_n)$ is mixing, but fails to be E_0-ergodic.

Consider the Cayley graph of $F_2 = \langle a, b\rangle$, i.e., the undirected graph with vertex set F_2 and edges between g, gh where $h \in \{a^{\pm 1}, b^{\pm 1}\}$. This is of course a tree. By a finite *subtree* (of this graph) we mean a finite subset $T \subseteq F_2$ such that T contains the unique path between any two of its elements (in particular, $\emptyset$ and $\{\gamma\}$ for $\gamma \in F_2$ are subtrees). For such a T and $s \in 2^T$, let

$$N_s = \{x \in 2^{F_2} : x|T = s\}.$$

Then an application of the usual generalization of the Kolmogorov Consistency Theorem (see [P, V.4.2]) shows that for any function $\varphi : \bigcup_T 2^T \to [0,1]$ which satisfies the following consistency condition, there is a unique measure μ on 2^{F_2} with $\mu(N_s) = \varphi(s)$:

Consistency condition. For any two finite subtrees T, T' with $T' = T \cup \{\gamma\}$, where $\gamma \notin T'$ and γ is connected by an edge to some vertex in T, we have, for $s \in 2^T$,

$$(*) \qquad \varphi(s) = \varphi(s\hat{\ }0) + \varphi(s\hat{\ }1),$$

where $s\hat{\ }i : T' \to 2$ is defined by $s\hat{\ }i|T = s, s\hat{\ }i(\gamma) = i$. Moreover for $T = \emptyset, \varphi(\emptyset) = 1$.

(Clearly every measure μ on 2^{F_2} arises in this fashion (take $\varphi(s) = \mu(N_s)$).)

To see this, take in the notation of Theorem V.4.2 of [P], $X = 2^{F_2}, \Delta = \{T : T$ is a finite subtree$\}$, $\leq \, = \, \subseteq, \mathcal{B}_T = \{\pi_T^{-1}(A) : A \subseteq 2^T\}$, where $\pi_T : 2^{F_2} \to 2^T$ is defined by $\pi_T(f) = f|T$, so that $\mathcal{B}_T$ is the σ-algebra generated by the sets $\{N_s : s \in 2^T\}$, $\mu_T(\pi_T^{-1}(A)) = \sum_{s\in A} \varphi(s)$. Condition (iv) in the statement of V.4.2 is trivially satisfied, since the atoms of $\mathcal{B}_T$ are exactly the sets $\{\pi_T^{-1}(\{s\}) : s \in 2^T\} = N_s$. Finally the system $\{\mu_T\}$ is consistent (according to the definition preceding V.4.2), as it can be easily verified using the condition $(*)$, noticing that if $S \leq T$ then there is a sequence $T_0 = S \leq T_1 \leq \cdots \leq T_n = T$, so that T_{i+1} is obtained from T_i by adding one vertex connected by an edge to some vertex of T_i.

Now fix a positive sequence $1 > p_n \to 0$, and for $s \in 2^T$, define $\varphi_n(s)$ as follows: If $T = \emptyset, \varphi_n(\emptyset) = 1$. If $T = \{\gamma\}$, and $s \in 2^T$, let $\varphi_n(s) = 1/2$.

Otherwise, given $s \in 2^T$, let K_0 be the number of edges $e = (\gamma_1, \gamma_2)$ with $\gamma_1, \gamma_2 \in T$, such that $s(\gamma_1) = s(\gamma_2)$ and K_1 be the number of edges, $e = (\gamma_1, \gamma_2)$ with $\gamma_1, \gamma_2 \in T$, where $s(\gamma_1) \neq s(\gamma_2)$. Put

$$\varphi_n(s) = \frac{1}{2}(1 - p_n)^{K_0} p_n^{K_1}.$$

Let μ_n be the corresponding measure, so that

$$\mu_n(N_s) = \frac{1}{2}(1 - p_n)^{K_0} p_n^{K_1}.$$

It is clear from the definition that μ_n is shift-invariant. Moreover, an argument similar to that of A6.1 shows that the action of F_2 on $(2^{F_2}, \mu_n)$ is mixing. Put

$$A = \{x \in 2^{F_2} : x(1) = 0\}.$$

Then, for any $\gamma \in F_2$,

$$A \triangle \gamma \cdot A = \{x \in 2^{F_2} : x(1) \neq x(\gamma)\}.$$

Assume now $\gamma \neq 1$. Let T_γ be the set of vertices in the unique path from 1 to γ. Now

$$A \triangle \gamma \cdot A \subseteq \bigcup \{N_s : s \in 2^{T_\gamma}, \text{ for some edge } e = (\gamma_1, \gamma_2) \text{ with } \gamma_1, \gamma_2 \in T_\gamma, \text{ we have } s(\gamma_1) \neq s(\gamma_2)\}.$$

So

$$\mu_n(A \triangle \gamma \cdot A) \leq M_\gamma p_n,$$

where M_γ is some positive constant depending on γ only, so $\mu_n(A \triangle \gamma \cdot A) \to 0$ as $n \to \infty$. It follows that if $A_n \subseteq \prod_n X_n$, where $X_n = 2^{F_2}$, is defined by

$$A_n = X_0 \times X_1 \times \cdots \times X_{n-1} \times A \times X_{n+1} \times \dots,$$

then for the diagonal action of F_2 on $\prod_n X_n$ and the measure $\mu = \prod_n \mu_n$, we have for any $\gamma \in F_2$, $\mu(A_n \triangle \gamma \cdot A_n) = \mu_n(A \triangle \gamma \cdot A) \to 0$ and

$$\mu(A_n) = \frac{1}{2},$$

so this action admits non-trivial almost invariant sets, thus it fails to be E_0-ergodic.

A7. Non-orbit equivalent relations

The theorem of Dye [D] together with the theorem of Ornstein-Weiss [OW], show that if Γ is an amenable group, any two Borel actions with non-atomic invariant, ergodic measure are OE. On the other hand we have:

Theorem A7.1 (Schmidt [S81]). *Let Γ be a non-amenable countable group. If Γ is not Kazhdan, then there are two free Borel actions of Γ with invariant, ergodic measure, which are not OE.*

Proof. Fix such a Γ. Since Γ is not amenable, the shift action of Γ on 2^Γ is E_0-ergodic by A4.1. Now since Γ is not Kazhdan, by A5.2 it admits a Borel action on a standard Borel space X with invariant, ergodic measure ν that is not E_0-ergodic. Fix a free mixing Borel action of Γ on a standard Borel space Y with invariant measure ρ (e.g., the free part of the shift action of Γ on 2^Γ) and consider the product space $X' = X \times Y$, with the measure $\mu' = \nu \times \rho$, and the diagonal action $\gamma \cdot (x, y) = (\gamma \cdot x, \gamma \cdot y)$. This is clearly free and μ' is invariant. The fact that μ' is ergodic was discussed in A6.

It is clear that $E_\Gamma^{X'}$ is not E_0-ergodic, therefore not OE to the shift action of Γ. $\dashv$

Open Problem A7.2. *Is it true that every non-amenable countable group has at least two free Borel actions with invariant, ergodic measure, that are not OE?*

Addendum. Hjorth has recently provided a positive answer to this question.

APPENDIX B

Cocycles and Cocycle-invariant Functions

B1. Review

We will recall here the basic definitions and some simple facts concerning cocycles. Detailed accounts can be found in Zimmer [Zi84] and Adams-Kechris [AK].

Let Γ be a countable group acting in a Borel way on a standard Borel space X with invariant measure μ, and H a countable group. A *cocycle* of this action into H is a Borel map $\alpha : \Gamma \times X \to H$ such that for all $\gamma, \delta \in \Gamma$,

$$\alpha(\gamma\delta, x) = \alpha(\gamma, \delta \cdot x)\alpha(\delta, x), \ \mu - \text{a.e. } (x).$$

If this equation is true for all x we say that α is a *strict cocycle.*

Two cocycles $\alpha : \Gamma \times X \to H, \beta : \Gamma \times X \to H$ are *equivalent* or *cohomologous*, in symbols

$$\alpha \sim \beta,$$

if there is a Borel map $x \mapsto h_x$ from X to H such that for all $\gamma \in \Gamma$,

$$\alpha(\gamma, x) = h_{\gamma \cdot x}\beta(\gamma, x)h_x^{-1}, \ \mu - \text{a.e. } (x).$$

A standard way in which cocycles arise is the following: Suppose H acts *freely* in a Borel way on the standard Borel space Y and $\rho : X \to Y$ is a Borel homomorphism of E_Γ^X to E_H^Y. Then define the (strict) cocycle $\alpha : \Gamma \times X \to H$ by

$$\alpha(\gamma, x) \cdot \rho(x) = \rho(\gamma \cdot x).$$

We call this the *cocycle associated* to the homomorphism ρ.

Let us note that if α is associated to ρ and $\beta \sim \alpha$, say $\alpha(\gamma, x) = h_{\gamma \cdot x}\beta(\gamma, x)h_x^{-1}$, μ-a.e. (x), then if we put

$$\sigma(x) = h_x^{-1} \cdot \rho(x),$$

σ is also a homomorphism of E_Γ^X to E_H^Y, such that moreover

$$\sigma(x) E_H^Y \rho(x), \text{ for all } x \in X,$$

and finally the cocycle associated to σ is equal to β, μ–a.e. In other words, we can "adjust" ρ to a homomorphism that only changes $\rho(x)$ up to E_H^Y-equivalence, to obtain any given cocycle equivalent to α.

Quite often in this paper we will be concerned about "cocycle reduction" results, which show that, under certain circumstances, a cocycle α can be replaced by an equivalent one β, $\alpha \sim \beta$, whose range $\beta(\Gamma \times X)$ is contained in a "small" subgroup of H.

B2. α-invariant functions

Suppose $\alpha : \Gamma \times X \to H$ is a Borel cocycle as in B1, and H acts in a Borel way on some space Y. Then a μ-measurable function $f : X \to Y$ is called *α-invariant* if for all $\gamma \in \Gamma$,

$$\alpha(\gamma, x) \cdot f(x) = f(\gamma \cdot x), \ \mu - \text{a.e. } (x)$$

We will often use the following variation of a basic fact, called the *Cocycle Reduction Lemma* in Zimmer [Zi84, 5.2.11].

Proposition B2.1. *Let a countable group Γ act in a Borel way on X with invariant measure μ and assume H is a countable group acting in a Borel way on Y, so that E_Γ^X is E_H^Y-ergodic. Let $\alpha : \Gamma \times X \to H$ be a Borel cocycle and assume that there is a μ-measurable, α-invariant function $f : X \to Y$. Then there is $y_0 \in Y$ and $\beta \sim \alpha$ with $\beta(\Gamma \times X) \subseteq H_{y_0}$ = the stabilizer of y_0.*

Proof. Since $\alpha(\gamma, x) \cdot f(x) = f(\gamma \cdot x)$, μ-a.e. (x), f restricted to a Borel co-null set X_0, is a homomorphism of $E_\Gamma^X | X_0$ to E_H^Y. So by the E_H^Y-ergodicity of E_Γ^X, there is $y_0 \in Y$ with $f(x) E_H^Y y_0$, μ-a.e. (x). Let $x \mapsto h_x$ be Borel such that

$$f(x) = h_x^{-1} \cdot y_0, \ \mu - \text{a.e. } (x).$$

Then note that if

$$\beta(\gamma, x) = h_{\gamma \cdot x} \alpha(\gamma, x) h_x^{-1},$$

$\beta \sim \alpha$, and

$$\beta(\gamma, x) \cdot y_0 = y_0, \ \mu - \text{a.e. } (x),$$

so $\beta(\gamma, x) \in H_{y_0}$, μ-a.e. (x). By defining $\beta(\gamma, x) = 1$ on a null set, we can actually assume that $\beta(\Gamma \times X) \subseteq H_{y_0}$. ⊣

B3. α-invariant measures

The following is a standard fact, which is explicitly isolated in Zimmer [Zi84, p. 78] but, in some form, traces back to Furstenberg [Fur]. Since we need to use it repeatedly, we include a proof below for completeness.

If K is a compact, metric space, we denote by $\mathcal{M}(K)$ the compact, metric space of measures on K (see, e.g., [Ke95, 17.E]). If a countable group H acts in a Borel way on K, then it acts in a Borel way on $\mathcal{M}(K)$ by

$$h \cdot \mu(A) = \mu(h^{-1} \cdot A),$$

for every Borel set $A \subseteq K$, or equivalently

$$\int f d(h \cdot \mu) = \int (h^{-1} \cdot f) d\mu,$$

where $h \cdot f(k) = f(h^{-1} \cdot k))$, for every (real or complex) continuous function f on K. If the action of H on K is continuous, so is the action of H on $\mathcal{M}(K)$.

Proposition B3.1 (see, e.g., Furstenberg [Fur], Zimmer [Zi84, p.78]). *Let Δ be a countable amenable group acting in a Borel way on a standard Borel space X with invariant measure μ. Let H be a countable group acting continuously on a compact, metric space K. Let $\alpha : \Delta \times X \to H$ be a Borel cocycle. Then there is an α-invariant, μ-measurable map $x \mapsto \nu_x$ from X to $\mathcal{M}(K)$.*

Proof. Let $\ell_1^\infty(X, \mathcal{M}(K))$ be the space of all measurable assignments

$$\lambda : X \to \mathcal{M}(K).$$

We equip this space with the topology provided by the semi-norms of the form

$$\lambda \mapsto \int_X (x \mapsto \int_K f_x d\lambda(x)) d\mu,$$

where $x \mapsto f_x$ is some measurable function from X to the continuous functions on K of norm ≤ 1.

The space $\ell_1^\infty(X, \mathcal{M}(K))$ can be thought of as a closed subset of the dual to the Banach space of measurable functions

$$x \mapsto f_x,$$

$$X \to C(K),$$

from X to the continuous functions on K, which have the property that

$$x \mapsto ||f_x||,$$

(where $\|f_x\| = \sup_{c \in K} |f_x(c)|$) is in $\ell^1(X, \mu)$. Thus in the topology generated by the seminorms it becomes a compact space, which is also convex, in the sense that if $\lambda_1, \lambda_2, \ldots, \lambda_n$ are in $\ell_1^\infty(X, \mathcal{M}(K))$ and $r_1, r_2, \ldots, r_n$ are non-negative reals summing to 1, then $\sum_{i \leq n} r_i \lambda_i \in \ell_1^\infty(X, \mathcal{M}(K))$.

We let Δ act on $\ell_1^\infty(X, \mathcal{M}(K))$ by

$$(h \cdot \lambda)(x) = \alpha(h, h^{-1} \cdot x) \cdot \lambda(h^{-1} \cdot x).$$

We let $(\Delta_n)_{n \in \mathbb{N}}$ be a Følner sequence for Δ. If we put

$$T_n : \ell_1^\infty(X, \mathcal{M}(K)) \to \ell_1^\infty(X, \mathcal{M}(K))$$

$$\lambda \mapsto \frac{1}{|\Delta_n|} \sum_{h \in \Delta_n} h \cdot \lambda$$

and take any point in the intersection of the closures of the images of the operators T_n, then we have a fixed point as required. ⊣

Under the assumption on CH one can obtain a stronger result.

Proposition B3.2 (Assume the Continuum Hypothesis). *Let Δ be a countable amenable group acting in a Borel way on a standard Borel space X with invariant measure μ. Let H be a countable group acting continuously on a compact, metric space K. Let $\alpha : \Delta \times X \to H$ be a Borel cocycle. Then there is an α-invariant, universally measurable map $x \mapsto \nu_x$ from X to $\mathcal{M}(K)$.*

Proof. Since Δ is amenable, a theorem of Christensen [C], Mokobodzki, gives, using CH, a universally measurable right-invariant mean on Δ, i.e., a positive linear functional φ on $\ell_\infty(\Delta)$, the Banach space of bounded real functions on Δ, with $\varphi(1) = 1$, and $\varphi(\gamma * p) = \varphi(p)$, where $\gamma * p(\delta) = p(\delta\gamma)$, which is such that $\varphi|[-1,1]^\Delta$ is a universally measurable map from $[-1,1]^\Delta$ into $[-1,1]$.

Now fix $\nu_0 \in \mathcal{M}(K)$ and define, for $x \in X$, a positive linear functional Λ_x on $C(K)$ = the Banach space of real continuous functions on K, by

$$\Lambda_x(f) = \varphi(\delta \mapsto \int f d(\alpha(\delta, x)^{-1} \cdot \nu_0)).$$

By the Riesz Representation Theorem, Λ_x corresponds to a measure $\nu_x \in \mathcal{M}(K)$, i.e.,

$$\Lambda_x(f) = \int f d\nu_x,$$

and the universal measurability of φ implies that $x \mapsto \nu_x$ is universally measurable. Finally, we verify that for every $\gamma \in \Delta$,

$$\alpha(\gamma, x) \cdot \nu_x = \nu_{\gamma \cdot x}, \ \mu\text{-a.e. } (x),$$

that is, for every $f \in C(K)$

$$\Lambda_x(\alpha(\gamma, x)^{-1} \cdot f) = \Lambda_{\gamma \cdot x}(f), \ \mu - \text{a.e. } (x)$$

or

$$\begin{aligned} &\varphi(\delta \mapsto \int (\alpha(\gamma, x)^{-1} \cdot f) d(\alpha(\delta, x)^{-1} \cdot \nu_0)) \\ &= \varphi(\delta \mapsto \int f d(\alpha(\delta, \gamma \cdot x)^{-1} \cdot \nu_0)), \ \mu - \text{a.e. } (x). \end{aligned}$$

Now

$$\alpha(\delta, \gamma \cdot x) = \alpha(\delta\gamma, x)\alpha(\gamma, x)^{-1}, \ \mu - \text{a.e. } (x),$$

so

$$\begin{aligned} &\int f d(\alpha(\delta, \gamma \cdot x)^{-1} \cdot \nu_0) \\ &= \int (\alpha(\delta, \gamma \cdot x) \cdot f) d\nu_0 \\ &= \int (\alpha(\delta\gamma, x)\alpha(\gamma, x)^{-1} \cdot f) d\nu_0, \ \mu - \text{a.e. } (x), \end{aligned}$$

and

$$\begin{aligned} &\int (\alpha(\gamma, x)^{-1} \cdot f) d(\alpha(\delta, x)^{-1} \cdot \nu_0) \\ &= \int (\alpha(\delta, x)\alpha(\gamma, x)^{-1} \cdot f) d\nu_0. \end{aligned}$$

Letting $p(\delta) = \int (\alpha(\delta, x)\alpha(\gamma \cdot x)^{-1} \cdot f) d\nu_0$, we have, by the right-invariance of φ,

$$\varphi(p) = \varphi(\gamma * p),$$

where $\gamma * p(\delta) = p(\delta\gamma)$, and the proof is complete. ⊣

Proposition B3.1 can be used to give a neat proof that free measure preserving actions of the free group on probability spaces are never hyperfinite (a special case of the fact mentioned at the end of 0F); and that neat proof can in turn be used to show that the analog of B3.1 fails in the Borel, or even Baire measurable, context.

Corollary B3.3. *If F_2 acts freely on a standard Borel space X with invariant measure μ, then $E^X_{F_2}$ is not hyperfinite.*

Proof. Appealing to the ergodic decomposition theorem, we may assume, without loss of generality, that F_2 acts ergodically. Write $F_2 = \langle a, b \rangle$, so that a and b are the generators. We suppose, towards a contradiction, that there is a Borel action of $\mathbb{Z}$ on X with $E^X_{\mathbb{Z}} = E^X_{F_2}$. (Recall that an equivalence relation is hyperfinite if and only if it is induced by a Borel action of $\mathbb{Z}$; see [DJK].)

We have a Borel cocycle

$$\alpha : \mathbb{Z} \times X \to F_2$$

given by

$$\alpha(k, x) = \sigma$$

if $k \cdot x = \sigma \cdot x$. Appealing to B3.1, and following the notation of C2, in using ∂F_2 to indicate the boundary of the Cayley graph of the free group, we may find an invariant conull Borel set $X_0 \subseteq X$ and a Borel assignment

$$x \mapsto \mu_x$$

$$X_0 \to \mathcal{M}(\partial F_2)$$

such that

$$\mu_{\ell \cdot x} = \alpha(\ell, x) \cdot \mu_x$$

for all $x \in X_0$, $\ell \in \mathbb{Z}$. Observing that for each x, $\sigma \in F_2$, there will be $\ell = \ell(\sigma, x) \in \mathbb{Z}$ with

$$\alpha(\ell, x) = \sigma,$$

we obtain

$$\mu_{\sigma \cdot x} = \sigma \cdot \mu_x,$$

for all $x \in X_0, \sigma \in F_2$.

Claim: $\mu_x \notin \mathcal{M}_3(\partial F_2)$, a.e. x.

(Here $\mathcal{M}_3(\partial F_2)$ refers to the collection of measures which are not supported on two or less points.)

Proof of claim: Otherwise ergodicity gives that for almost every $x, \mu_x \in \mathcal{M}_3(\partial F_2)$ and, appealing to lemma C2.3, we may find a Borel invariant conull set $X_0' \subseteq X_0$ and Borel function

$$s^* : X_0' \to F_2$$

such that for all $\sigma \in F_2, x \in X_0'$

$$s^*(\sigma \cdot x) = \sigma s^*(x).$$

Letting $A_1 = \{x \in X_0' : s^*(x) = 1\}$, we have that A_1 is a Borel transversal for $E_{F_2}^{X_0'}$, i.e., $E_{F_2}^{X_0'}$ is tame, which immediately contradicts the fact that F_2 acts freely with invariant measure on X_0'. (⊣Claim)

So we can assume that there is a Borel set $Y_0 \subseteq X_0$, conull and invariant, and a Borel function

$$Y_0 \to [\partial F_2]^{\leq 2}$$

$$x \mapsto \{e_x, e_x'\}$$

such that

$$\sigma \cdot \{e_x, e_x'\} = \{e_{\sigma \cdot x}, e_{\sigma \cdot x}'\},$$

for every $x \in Y_0, \sigma \in F_2$. (Here $[A]^{\leq 2}$ is the set of subsets of A of cardinality ≤ 2.)

Let A_0 be the set of $x \in Y_0$ such that neither e_x nor e_x' begin with a or a^{-1}. Here we view ∂F_2 and the action of F_2 on ∂F_2 as in the last paragraph of the proof of C2.1, i). For each $x \in Y_0$, we can find some $n \in \mathbb{Z}$ with $b^n \cdot x \in A_0$, thus $\mu(A_0) > 0$.

Also for $n_1 \neq n_2 \in \mathbb{Z}$,

$$a^{n_1} \cdot A_0 \cap a^{n_2} \cdot A_0 = \emptyset,$$

thus A_0 is a Borel transversal for $E_{\langle a \rangle}^{Z_0}$, where $\langle a \rangle = \{a^n : n \in \mathbb{Z}\}$, $Z_0 = \langle a \rangle \cdot A_0$, which is a contradiction as before. ⊣

Counterexample B3.4. *There is a continuous free action of $\mathbb{Z}$ on a Polish space X with dense orbits, and a Borel cocycle $\alpha : \mathbb{Z} \times X \to F_2$ such that there is* no *Baire measurable map*

$$X \to \mathcal{M}(\partial F_2),$$
$$x \mapsto \mu_x$$

such that for all x in a comeager invariant set, $n \in \mathbb{Z}$,

$$\mu_{n \cdot x} = \alpha(n, x) \cdot \mu_x.$$

One constructs the example as follows. Again, letting a and b be the generators of F_2, we start with the free part of the shift action of F_2 on 2^{F_2}; notice that this action is *generically ergodic*, in the sense that every invariant Borel set is either meager or comeager. Following [SWW], we may find an invariant dense G_δ set $X \subseteq 2^{F_2}$, contained in this free part, on which:

(a) $E^X_{F_2}$ is hyperfinite, and in fact there is a continuous action of $\mathbb{Z}$ on X with $E^X_{F_2} = E^X_{\mathbb{Z}}$.

(b) $\{a^\ell \cdot x : \ell \in \mathbb{Z}\}$ is dense in X, for all $x \in X$;

(c) $\{b^\ell \cdot x : \ell \in \mathbb{Z}\}$ is dense in X, for all $x \in X$.

It follows easily from (b) and (c) that the orbit equivalence relation $E^X_{F_2}$ as well as $E^X_{\langle a \rangle}$ and $E^X_{\langle b \rangle}$, induced by the respective cyclic subgroups $\langle a \rangle$ and $\langle b \rangle$, are generically ergodic, and hence non-tame (compare [Hj00a, 3.1]), and in fact non-tame even when restricted to any comeager subset of X. Let also $\alpha : X \times \mathbb{Z} \to F_2$ be the Borel cocycle defined as in the proof of B3.3.

Now, towards a contradiction, suppose that we can find a Baire measurable map $x \mapsto \mu_x$ with

$$\mu_{n \cdot x} = \alpha(n, x) \cdot \mu_x,$$

on an invariant comeager set X_0. As in the corollary above, we obtain

$$\mu_{\sigma \cdot x} = \sigma \cdot \mu_x,$$

for $x \in X_0, \sigma \in F_2$.

Claim: $\mu_x \notin \mathcal{M}_3(\partial F_2)$, for a comeager set of x.

Proof of claim: Otherwise generic ergodicity gives, on a comeager set $X_0' \subseteq X_0, \mu_x \in \mathcal{M}_3(\partial F_2)$, and, as before, this implies that $E^{X_0'}_{F_2}$ is tame, which is a contradiction. (⊣Claim)

So, again as before, on an invariant comeager Borel subset $Y_0 \subseteq X_0$, we can find a Borel map $x \mapsto \{e_x, e_x'\}$ with

$$\{e_{\sigma \cdot x}, e'_{\sigma \cdot x}\} = \sigma \cdot \{e_x, e_x'\}$$

for all $\sigma \in F_2$, $x \in Y_0$. We can again let A_0 be the set of $x \in Y_0$ for which neither e_x nor e_x' begins with an a or an a^{-1}. By considering the action of $\langle b \rangle$ we obtain that A_0 is non-meager. Thus $Z_0 = \langle a \rangle \cdot A_0$ is comeager, by generic ergodicity, and A_0 is a Borel transversal for $E^{Z_0}_{\langle a \rangle}$, contradicting its generic ergodicity.

Remark. Actually, the results of [SWW] imply the following: For each infinite countable group Γ and continuous free action of Γ on a perfect Polish space Y with dense orbits, there is an invariant dense G_δ set $Z \subseteq Y$, an invariant dense G_δ set $X \subseteq 2^{F_2}$ contained in the free part of 2^{F_2}, and a homeomorphism φ of Z onto X

with $xE^Z_\Gamma y \Leftrightarrow \varphi(x)E^X_{F_2}\varphi(y)$. In particular, the set $X \subseteq 2^{F_2}$ in B3.4 can be chosen so that there is a continuous free action of Γ on X with $E^X_{F_2} = E^X_\Gamma$. Thus, we can replace $\mathbb{Z}$ by any infinite countable group Γ in B3.4.

Finally, B3.4 can be also used to negatively answer a question of Weiss, see [W00], p. 290.

Fix an infinite countable group Γ and a free continuous action of Γ on a perfect Polish space X with dense orbits. Denote by $B(X)$ the linear space of all bounded Baire measurable functions $u : X \to \mathbb{R}$ modulo meager sets (i.e., the elements of $B(X)$ are equivalence classes of Baire measurable $f : X \to \mathbb{R}$ modulo the equivalence relation $f_1 \sim f_2 \Leftrightarrow f_1, f_2$ agree on a comeager set, and every $u \in B(X)$ is bounded modulo meager sets). Although technically a $u \in B(X)$ is an equivalence class, as usual we think of it as a function. Let also $U(X)$ be the linear space of all $\{u_n\} \in B(X)^{\mathbb{N}}$, which are uniformly bounded, i.e., for some c and all n, $|u_n(x)| < c$ on a comeager set of x's. Note that Γ acts on $B(X)$ by $(\gamma \cdot u)(x) = u(\gamma^{-1} \cdot x)$ and similarly on $U(X)$ by $\gamma \cdot \{u_n\} = \{\gamma \cdot u_n\}$.

A map $\pi : U(X) \to B(X)$ is called a *natural projection* if it satisfies:

(i) π is linear,

(ii) $\pi(\{u, u, \dots\}) = u$, for $u \in B(X)$,

(iii) $\underline{\lim}_n u_n \leq \pi(\{u_n\}) \leq \overline{\lim}_n u_n$, for $\{u_n\} \in U(X)$ (this of course means that for a comeager set of x's, $\underline{\lim}_n u_n(x) \leq \pi(\{u_n\}(x)) \leq \overline{\lim}_n u_n(x)$),

(iv) π is shift-invariant, i.e., $\pi(\{u_{n+1}\}) = \pi(\{u_n\})$, for $\{u_n\} \in U(X)$,

(v) π is a Γ-map, i.e., $\pi(\gamma \cdot \{u_n\}) = \gamma \cdot \pi(\{u_n\})$.

Note that, in the presence of the other conditions, (iii) is equivalent to

(iii)′ If $u_n \geq 0$ (i.e., $u_n(x) \geq 0$ on a comeager set) for all n, then $\pi(\{u_n\}) \geq 0$.

We now have the following result, which in the special case $\Gamma = \mathbb{Z}_2 \oplus \mathbb{Z}_2 \oplus \mathbb{Z}_2 \oplus \dots$ and the canonical action of Γ on $2^{\mathbb{N}}$ given by $((i_0, i_1, \dots) \cdot x)(k) = x(k) + i_k \bmod 2$, provides a negative answer to Weiss's question.

Theorem B3.5 *For any infinite amenable group Γ and continuous free action of Γ on a perfect Polish space X with dense orbits, there is no natural projection $\pi : U(X) \to B(X)$.*

Proof. In view of B3.4 and the remark following it, we can assume (by replacing X by an invariant dense G_δ set, if necessary), that there is a Borel cocycle $\alpha : \Gamma \times X \to F_2$ such that there is no Baire measurable map $X \to \mathcal{M}(\partial F_2), x \mapsto \mu_x$ with $\mu_{\gamma \cdot x} = \alpha(\gamma, x) \cdot \mu_x$, for all $\gamma \in \Gamma$, in a comeager set of x.

Assuming now that there is a natural projection $\pi : U(X) \to B(X)$, we will derive a contradiction, by finding such a map $x \mapsto \mu_x$.

Fix a (right) Følner sequence $\{F_n\}$ for Γ, i.e., a sequence of non-empty finite sets $F_n \subseteq \Gamma$ with $\frac{|F_n\gamma \Delta F_n|}{F_n} \to 0$, for all $\gamma \in \Gamma$. Fix also a measure $\nu_0 \in \mathcal{M}(\partial F_2)$, and put for $x \in X, n \in \mathbb{N}$

$$\nu_{n,x} = \frac{1}{|F_n|} \sum_{\delta \in F_n} \alpha(\delta, x)^{-1} \cdot \nu_0.$$

We view below $\mathcal{M}(\partial F_2)$ as a compact subset of the dual space $C(\partial F_2)^*$ of $C(\partial F_2)$, the Banach space of all continuous functions on ∂F_2. Norms refer to these spaces.

An easy calculation, using the cocycle identity, shows that for $\gamma \in \Gamma, x \in X$

$$||\nu_{n,\gamma\cdot x} - \alpha(\gamma, x) \cdot \nu_{n,x}|| \leq \frac{|F_n\gamma\Delta F_n|}{F_n} \to 0.$$

For $f \in C(\partial F_2)$, let $u_n^f : X \to \mathbb{R}$ be the Borel function defined by

$$u_n^f(x) = \nu_{n,x}(f).$$

and put

$$u^f = \pi(\{u_n^f\}).$$

(Note that $|u_n^f(x)| \leq ||f||$, for all x, n.)

Fix a countable $\mathbb{Q}$-subspace $\{f_i\} \subseteq C(\partial F_2)$, norm-dense in $C(\partial F_2)$, containing 1, and closed under the F_2-action on $C(\partial F_2)$. For any $x \in X$, define $\mu_x : \{f_i\} \to \mathbb{R}$ by

$$\mu_x(f_i) = u^{f_i}(x).$$

(Here we think of u^{f_i} as a function by picking a representative in its class.) Thus, $|\mu_x(f_i)| \leq ||f_i||$ on a comeager invariant set, using condition (iii) of a natural projection. It is also easy to check, using condition (i) of a natural projection, that μ_x is $\mathbb{Q}$-linear, on a comeager invariant set. It is positive ($f_i \geq 0 \Rightarrow \mu_x(f_i) \geq 0$) on a comeager invariant set, using condition (iii)$'$, and $\mu_x(1) = 1$, on a comeager invariant set, using condition (ii). Thus, μ_x extends uniquely to a measure, also denoted by μ_x, on a comeager invariant set. We let μ_x be equal to ν_0 outside this set, so $x \mapsto \mu_x$ is defined everywhere. Since the function $x \mapsto u^{f_i}(x)$ is Baire measurable for each i, it follows that $x \mapsto \mu_x$ is Baire measurable. Finally, to get a contradiction, it is enough to verify that on a comeager invariant set of x's,

$$\mu_{\gamma\cdot x} = \alpha(\gamma, x) \cdot \mu_x, \text{ for all } \gamma \in \Gamma.$$

This means that we need to check that for each $i, \gamma \in \Gamma$,

$$\mu_{\gamma\cdot x}(f_i) = (\alpha(\gamma, x) \cdot \mu_x)(f_i)$$

for a comeager set of x, or equivalently

$$\begin{aligned} u^{f_i}(\gamma \cdot x) &= \mu_x(\alpha(\gamma, x)^{-1} \cdot f_i) \\ &= u^{\alpha(\gamma,x)^{-1}\cdot f_i}(x) \end{aligned}$$

or

$$\pi(\{u_n^{f_i}\})(\gamma \cdot x) = \pi(\{u_n^{\alpha(\gamma,x)^{-1}\cdot f_i}\})(x).$$

Now, by condition (v),

$$\pi(\gamma^{-1} \cdot \{u_n^{f_i}\}) = \gamma^{-1} \cdot \pi(\{u_n^{f_i}\}),$$

so we have $\pi(\{u_n^{f_i}\})(\gamma \cdot x) = (\gamma^{-1} \cdot \pi(\{u_n^{f_i}\})(x) = \pi(\gamma^{-1} \cdot \{u_n^{f_i}\})(x)$, thus we only need to show that

$$\pi(\gamma^{-1} \cdot \{u_n^{f_i}\}) = \pi(\{u_n^{\alpha(\gamma,x)^{-1}\cdot f_i}\}).$$

Since conditions (i), (iii) imply that if $\{v_n\}, \{w_n\} \in U(X)$ are such that

$$|v_n(x) - w_n(x)| \to 0$$

for a comeager set of x, then

$$\pi(\{v_n\}) = \pi(\{w_n\}),$$

it is enough to verify that

$$|\gamma^{-1} \cdot u_n^{f_i}(x) - u_n^{\alpha(\gamma,x)^{-1}\cdot f_i}(x)| \to 0$$

on a comeager set of x, i.e.,

$$|\nu_{n,\gamma\cdot x}(f_i) - \nu_{n,x}(\alpha(\gamma,x)^{-1}\cdot f_i)| \to 0$$

or equivalently

$$|\nu_{n,\gamma\cdot x}(f_i) - (\alpha(\gamma,x)\cdot\nu_{n,x})(f_i)| \to 0,$$

which is immediate, since

$$||\nu_{n,\gamma\cdot x} - \alpha(\gamma,x)\cdot\nu_{n,x}|| \to 0,$$

and the proof is complete. ⊣

Remark. Note that condition (iv) of a natural projection was never used in this proof. However, if one has a π that satisfies all the other conditions except (iv), then $\pi'(\{u_n\}) = \pi(\{\frac{1}{n+1}\sum_{i\leq n} u_n\})$ satisfies all conditions (i)-(v).

Finally, note that if "Baire measurable" is replaced by "measurable" (with respect to some fixed measure), then, by the Christensen, Mokobodzki result, using CH, there is a natural projection from uniformly bounded sequences of measurable functions into bounded measurable functions.

B4. Maximum two-supported measures

We will prove here a basic fact, which goes back, in some form or another, to papers of Adams and Zimmer, e.g., [Zi81, 3.7], [Ad88, 3.1], [Ad95, 2.6]. We will make frequent use of this and some of its variations in this paper.

We use the following notation below: If K is a compact, metric space, we denote by $\mathcal{M}_{\leq 2}(K)$ the Borel subset of $\mathcal{M}(K)$ consisting of measures supported by at most 2 points, i.e., the $\nu \in \mathcal{M}(K)$ for which there are $a, b \in K$ (not necessarily distinct) with $\nu(\{a,b\}) = 1$. Put $\text{supp}(\nu) = \{a,b\}$ in this case. If H acts continuously on K and we consider the induced action on $\mathcal{M}(K)$, the set $\mathcal{M}_{\leq 2}(K)$ is clearly invariant. Put also $\mathcal{M}_3(K) = \mathcal{M}(K)\setminus\mathcal{M}_{\leq 2}(K)$.

Proposition B4.1 (Adams, Zimmer). *Let Γ be a countable group acting in a Borel way on a standard Borel space X with invariant measure μ. Let H be a countable group acting in a continuous way on a compact, metric space K and consider the induced action on $\mathcal{M}(K)$. Let $\alpha : \Gamma\times X \to H$ be a Borel cocycle. Let $\mathcal{S}$ denote the set of all μ-measurable, α-invariant functions $x \mapsto \nu_x$ from X to $\mathcal{M}(K)$, and assume that $\mathcal{S} \neq \emptyset$ and all $x \mapsto \nu_x$ in $\mathcal{S}$ satisfy $\nu_x \in \mathcal{M}_{\leq 2}(K)$, μ-a.e. (x). Define a partial pre-order $\precsim$ on $\mathcal{S}$ by letting, for $\nu_1, \nu_2 \in \mathcal{M}_{\leq 2}(K)$,*

$$\nu_1 \leq \nu_2 \Leftrightarrow \text{supp}(\nu_1) \subseteq \text{supp}(\nu_2)$$

and

$$(x\mapsto\nu_x) \precsim (x\mapsto\nu'_x) \text{ iff } (\nu_x \leq \nu'_x,\ \mu\text{-a.e. } (x)).$$

Then there is a maximum element in $(\mathcal{S},\precsim)$.

Proof. Consider any $S \in \mathcal{S}$, where for $S = (x\mapsto\nu_x)$ we write $S(x) = \nu_x$. Put

$$D(S) = \{x \in X : |\text{supp}(\nu_x)| = 2\},$$

with $|A| = \text{card}(A)$. Then $D(S)$ is μ-measurable and we put

$$r = \sup\{\mu(D(S)) : S \in \mathcal{S}\}.$$

We first argue that this sup is attained. Indeed, pick $S_n = (x\mapsto\nu_x^n) \in \mathcal{S}$ with $\mu(D(S_n)) > r - \frac{1}{n}$. Let $\bar{\nu}_x^n = \frac{1}{2}(\delta_{a^n(x)} + \delta_{b^n(x)})$, where $\text{supp}(\nu_x^n) = \{a^n(x), b^n(x)\}$.

Here δ_a = the Dirac measure at a. Then clearly $\bar{S}_n = (x \mapsto \bar{\nu}_x^n) \in \mathcal{S}$ as well. Put $S_\infty = (x \mapsto \nu_x^\infty)$, where

$$\nu_x^\infty = \sum 2^{-n} \nu_x^n.$$

Then again $S_\infty \in \mathcal{S}$, so $\nu_x^\infty \in \mathcal{M}_{\leq 2}(K)$, μ-a.e. (x), by our assumption. Clearly $D(S_\infty) \supseteq D(\bar{S}_n) = D(S_n)$, for each n, so $\mu(D(S_\infty)) = r$.

We will now see that actually S_∞ is a maximum element of $(\mathcal{S}, \precsim)$. Indeed, fix any $S = (x \mapsto \nu_x) \in \mathcal{S}$. By a similar argument to the above, $\frac{(S+S_\infty)}{2} \in \mathcal{S}$, so $\mu(D(S) \setminus D(S_\infty)) = 0$, since, otherwise, $\mu(D(\frac{S+S_\infty}{2})) > \mu(D(S_\infty)) = r$. Also for the same reason,

$$\begin{aligned} \mu(\{x \in X : &|\mathrm{supp}(\nu_x)| = 1, |\mathrm{supp}(\nu_x^\infty)| = 1, \\ &\mathrm{supp}(\nu_x) \not\subseteq \mathrm{supp}(\nu_x^\infty)\} = 0, \end{aligned}$$

and, by our assumption on $\mathcal{S}$,

$$\begin{aligned} \mu(\{x \in X : &|\mathrm{supp}(\nu_x)| = 1, |\mathrm{supp}(\nu_x^\infty) = 2|, \\ &\mathrm{supp}(\nu_x) \not\subseteq \mathrm{supp}(\nu_x^\infty)\} = 0, \end{aligned}$$

It then follows that $\mathrm{supp}(\nu_x) \subseteq \mathrm{supp}(\nu_x^\infty)$, μ-a.e. (x), i.e., $\nu_x \leq \nu_x^\infty$, μ-a.e. (x), or $S \preceq S_\infty$. ⊣

APPENDIX C

Actions on Boundaries

C1. Trees

In this paper, by a *tree* we will always mean an acyclic connected graph $\langle V, T\rangle$, with countable vertex set V, and edge relation T (i.e., $T \subseteq V^2$ and $(x,y) \in T$ iff x, y are connected by an edge). When V is understood or irrelevant, we simply write T for the tree. A *finite path* in T is a sequence $(v_0, v_1, \dots, v_n)$ with $n \geq 1, (v_i, v_{i+1}) \in T$, and $v_i \neq v_j$ for $i \neq j$.

The boundary of the tree T, in symbols ∂T, is defined as follows:

An (infinite) path through T is a sequence $(v_0, v_1, \dots)$ such that $v_i \neq v_j$ for $i \neq j$, and $(v_i, v_{i+1}) \in T$ for each i. We call two paths $(v_n), (w_n)$ equivalent, in symbols, $(v_n) \sim (w_n)$, if $\exists n \exists m \forall i (v_{n+i} = w_{m+i})$. An *end* of T is an equivalence class of paths. The *boundary* of T is the set of ends of T.

It will be also convenient to use another (but closely related) concept of tree, which is a standard tool in descriptive set theory. To distinguish it from the notions above, we will call it a set-theoretic tree.

A *set-theoretic tree* on a countable set A is a subset $S \subseteq A^{<\infty}$, the set of finite sequences from S, such that $\emptyset$ (=the empty sequence) is in S and if $(a_0, \dots, a_{n-1}) \in S$ and $m \leq n$, then $(a_0, \dots, a_{m-1}) \in S$. The *body* of S is the set

$$[S] = \{(a_0, a_1, \dots) \in A^{\mathbb{N}} : \forall n (a_0, \dots, a_{n-1}) \in S\}.$$

When $A^{\mathbb{N}}$ is equipped with the product topology, with A discrete, $[S]$ is a closed subspace of $A^{\mathbb{N}}$. A basis for the topology of $[S]$ is given by the sets $N_{(a_0,\dots,a_{n-1})} = \{(b_0, b_1, \dots) \in [S] : \forall i < n (b_i = a_i)\}$, for $(a_0, \dots, a_{n-1}) \in S$.

Now suppose (V, T) is a tree and fix $v_0 \in V$. Then for each end $e \in \partial T$ there is a unique path $(v_0, v_1, \dots) \in e$ starting from v_0, called the *geodesic* from v_0 to e and denoted by $[v_0, e]$. Let

$$T(v_0) = \{\emptyset\} \cup \{(v_1, v_2, \dots, v_{n-1}) : n \geq 2, (v_0, v_1, v_2, \dots, v_{n-1}) \text{ is a finite path of } T\}.$$

Then $T(v_0)$ is a set-theoretic tree on V, and so $[T(v_0)]$ is a closed subset of $V^{\mathbb{N}}$. The map $\varphi_{v_0} : [T(v_0)] \to \partial T$ given by

$$\varphi_{v_0}(x) = e \text{ iff } x = [v_0, e],$$

is a bijection for $[T(v_0)]$ onto ∂T, and can be used to put a topology on ∂T, by transferring, via this map, the topology of $[T(v_0)]$. It is important to note here that this topology is independent of v_0, so intrinsically defined, since if w_0 is another vertex, the map $\varphi_{w_0}^{-1} \circ \varphi_{v_0} : T(v_0) \to T(w_0)$ is a homeomorphism. A basis for the topology of ∂T is given by the sets

$$[v_0, \dots, v_n] = \{e \in \partial T : \exists v_{n+1}, v_{n+2}, \dots \\ (v_0, v_1, v_2, \dots, v_n, v_{n+1}, \dots) \in e\},$$

with $(v_0, v_1, \dots, v_n)$ a finite path in T.

It is clear that if T is *locally finite*, i.e., every vertex has finite degree, ∂T is compact, metrizable.

When T is not necessarily locally finite, we can define a canonical compactification of ∂T as follows:

Let $\partial^* T = \partial T \cup V$. We now define the topology on $\partial^* T$. Let $(v_0, v_1, \dots, v_n)$ be a finite path in T. We allow the possibility that $n = 0$, i.e., a single vertex (v_0). First define $[v_0, v_1, \dots, v_n]^*$ for $n \geq 1$, by

$$[v_0, \dots, v_n]^* = [v_0, \dots, v_n] \cup \{v \in V : \\ \exists v_{n+1}, \dots, v_m[(v_0, v_1, \dots, v_n, v_{n+1}, \dots v_m = v) \\ \text{is the unique path from } v_0 \text{ to } v]\}$$

Next, for $v_0 \in V$ and F a finite set of vertices, let

$$[v_0]^*_F = \{v_0\} \cup \bigcup \{[v_0, v_1]^* : (v_0, v_1) \in T, v_1 \notin F\}.$$

Finally, the basis for the topology of $\partial^* T$ consists of all the sets

$$[v_0, v_1, \dots, v_n]^*, [v_0]^*_F,$$

for $n \geq 1, (v_0, \dots, v_n)$ a path in T, $v_0 \in V$, F a finite set of vertices.

It is not hard to check that $\partial^* T$ is compact, metrizable and ∂T is a subspace of $\partial^* T$.

Similarly, if $S \subseteq A^{<\infty}$ is a set theoretic tree on A, we define the compactification $[S]^*$ of S by letting

$$[S]^* = [S] \cup S,$$

and taking as a basis for the topology of $[S]^*$ the sets

$$N^*_{(a_0, \dots, a_{n-1})} = N_{(a_0, \dots, a_{n-1})} \cup \{(b_0, \dots, b_{m-1}) : \\ m \geq n, \forall i < n (b_i = a_i)\}$$

for $(a_0, \dots, a_{n-1}) \in S$, and

$$N^*_{a_0, \dots, a_{n-1}, F} = \{(a_0, \dots, a_{n-1})\} \cup \{N^*_{(a_0, \dots, a_{n-1}, a_n)} : \\ (a_0, \dots, a_{n-1}, a_n) \in S, a_n \notin F\}$$

for $(a_0, \dots, a_{n-1}) \in S, F \subseteq A$ finite.

Then, if (V, T) is a tree and $v_0 \in V$, the map

$$\varphi^*_{v_0} : [T(v_0)]^* \to \partial^* T$$

given by

$$\varphi^*_{v_0} | [T(v_0)] = \varphi_{v_0}, \\ \varphi^*_{v_0}(\emptyset) = v_0 \\ \varphi^*_{v_0}(v_1, \dots, v_{n-1}) = v_{n-1}, \text{ for } n \geq 2$$

is a homeomorphism of $[T(v_0)]^*$ with $\partial^* T$.

For each set X, we denote by $[X]^k$ the set of subsets of X of cardinality k ($k = 0, 1, 2, \dots$). Let also $[X]^{<\infty} = \bigcup_k [X]^k$. If T is a tree and $\{e_1, e_2\} \in [\partial T]^2$, then the (*geodesic*) line from e_1 to e_2 in symbols

$$[e_1, e_2]$$

is a sequence $(v_n)_{n\in\mathbb{Z}}$ such that $(v_n, v_{n+1}) \in T, \forall n \in \mathbb{Z}$, and

$$[v_0, e_2] = (v_0, v_1, v_2, \dots), [v_0, e_1] = (v_0, v_{-1}, v_{-2}, \dots).$$

It is uniquely determined up to a shift. We will also occasionally write $[e_1, e_2] = \{v_n : n \in \mathbb{Z}\}$ for the set of vertices on this line. Finally, if $\{e_1, e_2, e_3\} \in [\partial T]^3$, then there is a unique vertex in V, denoted by

$$[e_1, e_2, e_3],$$

which belongs to $[e_1, e_2], [e_2, e_3], [e_3, e_1]$.

Suppose now Γ is a countable group acting by automorphisms on T, i.e., Γ acts on V and $(x, y) \in T \Leftrightarrow (\gamma \cdot x, \gamma \cdot y) \in T$. We will simply say in this case that Γ *acts on* T. Since $(x_i) \sim (y_i) \Leftrightarrow (\gamma \cdot x_i) \sim (\gamma \cdot y_i)$, Γ acts naturally on ∂T by

$$\gamma \cdot [(x_i)]_\sim = [(\gamma \cdot x_i)]_\sim.$$

This action is continuous. Similarly, Γ acts continuously on $\partial^* T$. Of course Γ acts in a natural (Borel) way on the standard Borel spaces $[\partial T]^2, [\partial T]^3$ as well, and the map $\{e_1, e_2, e_3\} \mapsto [e_1, e_2, e_3]$ is a Borel Γ-map from $[\partial T]^3$ to T.

C2. Free groups

Consider now $F_n, 1 \leq n \leq \infty$, the free group of rank n and fix a set of generators $a_1, a_2, \dots$ for F_n. The *Cayley graph* of F_n (relative to this fix set of generators) has vertex set $V = F_n$ and edge relation T given by

$$(\gamma, \delta) \in T \Leftrightarrow \exists i(\delta = \gamma a_i \text{ or } \gamma = \delta a_i).$$

The group F_n acts on T by left-multiplication

$$\gamma \cdot \delta = \gamma\delta.$$

We will consider now the action of F_n on ∂T, which is usually denoted by ∂F_n and called the *boundary* of F_n. The following results are part of the folklore.

Proposition C2.1. *i) The action of F_n on ∂F_n has hyperfinite associated equivalence relation $E_{F_n}^{\partial F_n}$. Similarly the action of F_n on $[\partial F_n]^2$ has hyperfinite associated equivalence relation $E_{F_n}^{[\partial F_n]^2}$.*

ii) For $e \in \partial F_n$, the stabilizer of e is either trivial or isomorphic to $\mathbb{Z}$. There are only countably many $e \in \partial F_n$ for which the stabilizer is non-trivial. Similarly for $\{e_1, e_2\} \in [\partial F_n]^2$, the stabilizer of $\{e_1, e_2\}$ is either trivial or isomorphic to $\mathbb{Z}$ and there are only countably many $\{e_1, e_2\}$ for which it is non-trivial.

Proof. i) Note that the first assertion implies the second. To see this, fix a Borel ordering $\prec$ of ∂F_n, let $X = \{(e_1, e_2) \in (\partial F_n)^2 : e_1 \neq e_2\}$ and define the involution π on X by $\pi((e_1, e_2)) = (e_2, e_1)$. Identify $[\partial F_n]^2$ with the Borel subset $X_0 = \{(e_1, e_2) \in (\partial F_n)^2 : e_1 \prec e_2\}$. If $E_{F_n}^{\partial F_n}$ is hyperfinite, so is $E = (E_{F_n}^{\partial F_n} \times E_{F_n}^{\partial F_n})|X$, and thus so is E' on X defined by

$$xE'y \Leftrightarrow xEy \text{ or } xE\pi(y),$$

since E' has finite index over $E|X$; see [JKL, 1.3]. Now clearly $E_{F_n}^{[\partial F_n]^2} \subseteq E'$, so $E_{F_n}^{[\partial F_n]^2}$ is hyperfinite.

To prove that $E_{F_n}^{\partial F_n}$ is hyperfinite, choose $v_0 = 1$ (the identity of F_n) as a fixed vertex in F_n. Then given $e \in \partial F_n$, the geodesic $[1, e]$ is simply an infinite reduced word $s_0 s_1 s_2 \dots$ in the generators $a_1, a_2, \dots$, i.e., an infinite sequence (s_i) such that s_i is of the form $a_j^{\pm 1}$ and $s_i s_{i+1} \neq 1$. Thus we can identify ∂F_n with the set of such words, which is a closed subset of $\{a_1, a_2, \dots\}^{\mathbb{N}}$. With this identification, it is easy to check that the action of $\gamma = t_0 \dots t_k \in F_n$ on $s_0 s_1 s_2 \dots$, where $t_0 \dots t_k$ is a finite reduced word, is given by left-concatenating

$$t_0 t_1 \dots t_k s_0 s_1 s_2 \dots$$

and then doing the obvious cancelations. It follows that, under this identification, the equivalence relation $E_{F_n}^{\partial F_n}$ corresponds to the following *tail* equivalence relation on the set of infinite reduced words:

$$(s_i) E_t (t_i) \Leftrightarrow \exists n \exists m \forall i (s_{n+i} = t_{m+i}),$$

which by [DJK, 8.2] is hyperfinite, and the proof of i) is complete.

ii) Using the notation and concepts of i), it is easy to check that if an infinite reduced word has non-trivial stabilizer, then it is of the form

$$e = w\hat{\ } p\hat{\ } p\hat{\ } \dots$$

where $w, p \in F_n$. So clearly there are only countably many such e. Replacing e by $w^{-1} \cdot e$ we get $p\hat{\ } p\hat{\ } p\hat{\ } \dots$, whose stabilizer is a conjugate of the stabilizer of e, so we may as well assume that $e = p\hat{\ } p\hat{\ } \dots$ with p of least possible length. Then it is easy to check that its stabilizer is $\{p^n : n \in \mathbb{Z}\}$.

Now suppose $\{e_1, e_2\} \in [\partial F_n]^2$. If $1 \neq \gamma \in \mathrm{Stab}(\{e_1, e_2\})$, then clearly $\gamma^2 \in \mathrm{Stab}(e_1) \cap \mathrm{Stab}(e_2)$. It follows that there are only countably many $\{e_1, e_2\}$ whose stabilizer is non-trivial. For each $\{e_1, e_2\}$, if γ is as above and $\mathrm{Stab}(e_1) = \{\delta^n : n \in \mathbb{Z}\}$, then $\delta^n = \gamma^2$ for some $n \neq 1$, so since commutativity of non-trivial elements is an equivalence relation in any free group (see, e.g. [LS, 2.18]) it follows that γ, δ commute. Thus the stabilizer of $\{e_1, e_2\}$ is abelian, so isomorphic to $\mathbb{Z}$. ⊣

Finally, we consider the action of F_n on $\mathcal{M}(\partial F_n)$, for finite n (so that the spaces $\partial F_n, \mathcal{M}(\partial F_n)$ are compact).

Proposition C2.2. *Let n be finite, and consider the action of F_n on $\mathcal{M}(\partial F_n)$.*

i) The action of F_n on $\mathcal{M}_3(\partial F_n)$ is free and the corresponding equivalence relation $E_{F_n}^{\mathcal{M}_3(\partial F_n)}$ is tame.

ii) There is a Borel partition $\mathcal{M}_{\leq 2}(\partial F_n) = \mathcal{M}_1 \cup \mathcal{M}_2$ into F_n-invariant Borel sets, such that the action of F_n on $\mathcal{M}_1$ is free and $E_{F_n}^{\mathcal{M}_1}$ is hyperfinite, and the action of F_n on $\mathcal{M}_2$ has stabilizers isomorphic to $\mathbb{Z}$, and $F_{F_n}^{\mathcal{M}_2}$ is tame.

Proof. i) This is clear from the following lemma, where F_n acts on itself by left-translation ($\gamma \cdot \delta = \gamma\delta$).

Lemma C2.3. *There is a Borel F_n-map from $\mathcal{M}_3(\partial F_n)$ to F_n.*

Proof (Lyons, see [AL]). Fix $\mu \in \mathcal{M}_3(\partial F_n)$. Using Fubini (see, e.g., [JKL, proof of 2.24]), if μ^3 is the product of three copies of μ (a measure on $(\partial F_n)^3$), then

$\mu^3([\partial F_n]^3) > 0$, where we write here $[\partial F_n]^3$ for the set of triples $(x, y, z) \in (\partial F_n)^3$ with x, y, z distinct. Consider the Borel F_n-map

$$\varphi(\{e_1, e_2, e_3\}) = [e_1, e_2, e_3],$$

and let $\nu = \varphi_*((\mu^3|[\partial F_n]^3)/\mu^3([\partial F_n]^3))$. This is a measure on F_n, so let

$$\theta(\mu) = \{\gamma \in \partial F_n : \nu(\{\gamma\}) \text{ is maximum}\}.$$

Clearly $\theta(\mu) \in [F_n]_0^{<\infty}$, the set of finite non-empty subsets of F_n, and if F_n acts on $[F_n]_0^{<\infty}$ in the obvious way by left-translation, then clearly θ is a Borel F_n-map from $\mathcal{M}_3(\partial F_n)$ to $[F_n]_0^{<\infty}$. Now, since F_n is torsion free, F_n acts freely on $[F_n]_0^{<\infty}$, so there is an F_n-map $\eta : [F_n]_0^{<\infty} \to F_n$. Then

$$\eta \circ \theta$$

is a Borel F_n-map from $\mathcal{M}_3(\partial F_n)$ to F_n. ⊣

ii) Decompose $\mathcal{M}_{\leq 2}(\partial F_n)$ as $\mathcal{M}_{\leq 2}(\partial F_n) = \mathcal{M}_0 \sqcup \bigsqcup_{0<r<\frac{1}{2}} \mathcal{M}_r \sqcup \mathcal{M}_{\frac{1}{2}}$, where $\mathcal{M}_0$ consists of all measures supported by exactly one point, $\mathcal{M}_{\frac{1}{2}}$ consists of all measures supported by two points, each one having measure $\frac{1}{2}$, and $\mathcal{M}_r$ consists of all measures of the form $r\delta_a + (1-r)\delta_b, a \neq b$. Then, by the proof of C2.1, it is easy to see that each one of the sets $\mathcal{M}_r$ $(0 \leq r \leq \frac{1}{2})$ can be split into two F_n-invariant Borel sets $\mathcal{M}_r^1, \mathcal{M}_r^2$ such that the action of F_n on $\mathcal{M}_r^1$ is free and $F_{F_n}^{\mathcal{M}_r^1}$ is hyperfinite, and the action of F_n on $\mathcal{M}_r^2$ has stabilizers isomorphic to $\mathbb{Z}$ and $\mathcal{M}_r^2$ is countable. Put then

$$\mathcal{M}_1 = \bigcup_{0 \leq r \leq \frac{1}{2}} \mathcal{M}_r^1$$

$$\mathcal{M}_2 = \bigcup_{0 \leq r \leq \frac{1}{2}} \mathcal{M}_r^2.$$

⊣

Corollary C2.4. *The groups $F_n, 1 \leq n \leq \infty$, are near-hyperbolic.*

Proof. This is clear from C2.2, if n is finite, and it follows for F_∞, since $F_\infty \subseteq F_2$. ⊣

C3. Free products of finite groups

We will consider here free products

$$\Gamma = A_0 * A_1 * \ldots,$$

of finite non-trivial groups, where the sequence $A_0, A_1, \ldots$ may be finite or infinite. The facts below are part of the folklore.

We associate to Γ the following set theoretic tree S_Γ on $A = \bigsqcup_i (A_i \setminus \{1\})$:

$$(a_0, a_1, \ldots, a_{n-1}) \in S_\Gamma \Leftrightarrow a_0 a_1 \ldots a_{n-1} \text{ is a reduced word in } \Gamma,$$

where $a_0 a_1 \ldots a_{n-1}$ is *reduced* if $a_i \in A_{n_i} \setminus \{1\}$ with $n_i \neq n_{i+1}, \forall i \leq n-2$. Thus the elements of S_Γ are in 1-1 correspondence with the elements of Γ:

$$(a_0, \ldots, a_{n-1}) \leftrightarrow a_0 \ldots a_{n-1}, \text{ for } n \geq 1,$$

$$\emptyset \leftrightarrow 1.$$

The infinite sequences $(a_0, a_1, \ldots) \in [S_\Gamma]$ can be viewed as infinite reduced words.

The group Γ acts on S_Γ by left-concatenation and cancelation, i.e., for

$$(a_0, a_1, \dots, a_{n-1}) \in S_\Gamma, \gamma \in \Gamma,$$

$$\gamma \cdot (a_0, \dots, a_{n-1}) = \text{the reduced work equal to } \gamma a_0 \dots a_{n-1}.$$

Similarly Γ acts continuously on $[S_\Gamma]$ by left-concatenation and cancelation and also acts continuously on $[S_\Gamma]^* = [S_\Gamma] \cup S_\Gamma$. We will now prove results analogous to C2 for Γ.

Proposition C3.1. *i) The action of Γ on $[S_\Gamma]$ has hyperfinite associated equivalence relation $E_\Gamma^{[S_\Gamma]}$. Similarly the action of Γ on $[S_\Gamma]^2$ has hyperfinite associated equivalence relation $E_\Gamma^{[S_\Gamma]^2}$.*

ii) For $e \in [S_\Gamma]$, the stabilizer of e is cyclic. There are only countably many $e \in [S_\Gamma]$ for which the stabilizer is non-trivial. Similarly for $\{e_1, e_2\} \in [S_\Gamma]^2$, the stabilizer of $\{e_1, e_2\}$ is cyclic-by-finite and there are only countably many $\{e_1, e_2\}$ for which it is non-trivial.

Proof. i) The proof here is essentially identical to that of C2.1, i), since the equivalence relation $E_\Gamma^{[S_\Gamma]}$ is simply the tail equivalence relation.

ii) The part concerning $[S_\Gamma]$ is again similar to that of C2.1, ii). Concerning $[S_\Gamma]^2$, we have again that if $\{e_1, e_2\} \in [S_\Gamma]^2$ and $1 \neq \gamma \in \text{Stab}(\{e_1, e_2\})$, then $\gamma^2 \in \text{Stab}(\{e_1\}) \cap \text{Stab}(\{e_2\})$. It follows that $\text{Stab}(\{e_1, e_2\})$ cannot contain a free non-abelian subgroup. Now, by Kurosh's Theorem (see, e.g., [Ro, 11.55]) any subgroup of Γ is isomorphic to a free product of a free group (perhaps trivial) and subgroups of each A_i. Since the free product $A * B$ of two groups with $|A| \geq 3, |B| \geq 2$ contains a free non-abelian subgroup (see, e.g., [LS, p. 177]), it follows that each stabilizer must be cyclic-by-finite. ⊣

Let us also note the following fact concerning the action of Γ on $[S_\Gamma]^3$.

Proposition C3.2. *There is a Borel Γ-map $\varphi : [S_\Gamma]^3 \to [\Gamma]^{<\infty}$.*

Proof. For $e \neq f \in [S_\Gamma]$, let $e \wedge f =$ the largest common initial segment of $e, f \in [S_\Gamma]$. For $\{e, f, g\} \in [S_\Gamma]^3$, the sequences $e \wedge f, f \wedge g, g \wedge e$ are compatible, i.e., each for each two of them one is an initial segment of the other, so let $\langle e, f, g \rangle$ be the union of $e \wedge f, f \wedge g, g \wedge e$, i.e., the longest of these three sequences. Say

$$\langle e, f, g \rangle = a_0 a_1 \dots a_{n-1},$$

where $n \geq 0$ (so that $\langle e, f, g \rangle$ could be $\emptyset$). Consider now two cases:

(a) $\langle e, f, g \rangle = e \wedge f = f \wedge g = g \wedge e$. Then $e = a_0 a_1 a_2 \dots a_{n-1} e_n \dots, f = a_0 a_1 \dots a_{n-1} f_n \dots, g = a_0 a_1 \dots a_{n-1} g_n \dots$, and say $e_n \in A_{k_n}, f_n \in A_{\ell_n}, g_n \in A_{m_n}$. If k_n, ℓ_n, m_n are distinct, put

$$\varphi(\{e, f, g\}) = \langle e, f, g \rangle = \{a_0 \dots a_{n-1}\}.$$

If at least two of k_n, ℓ_n, m_n are equal, with common value k, put

$$\varphi(\{e, f, g\}) = a_0 \dots a_{n-1} A_k.$$

(b) One of $e \wedge f, f \wedge g, g \wedge e$ is longer than the other two, say, e.g., $e \wedge f$. Then $e = a_0 a_1 \dots a_{n-1} e_n \dots, f = a_0 a_1 \dots a_{n-1} f_n \dots$, and say $e_n \in A_{k_n}, f_n \in A_{\ell_n}$. If $k_n \neq \ell_n$, let

$$\varphi(\{e, f, g\}) = \langle e, f, g \rangle = \{a_0 \dots a_{n-1}\}.$$

If $k_n = \ell_n = k$, put

$$\varphi(\{e, f, g\}) = a_0 \dots a_{n-1} A_k.$$

It is easy to check that $\varphi : [S_\Gamma]^3 \to [\Gamma]^{<\infty}$ is a Γ-map. ⊣

Finally we verify that Γ is near-hyperbolic.

Proposition C3.3. *Let $\Gamma = A_0 * A_1 * \ldots$ be a free product of finite groups. Then Γ is near-hyperbolic.*

Proof. When there are only finitely many factors, $[S_\Gamma]$ is compact, so consider the continuous action of Γ on $[S_\Gamma]$ and the induced action on $\mathcal{M}([S_\Gamma])$.

The action of Γ on $\mathcal{M}_{\leq 2}([S_\Gamma])$ has cyclic-by-finite stabilizers and $E_\Gamma^{\mathcal{M}_{\leq 2}([S_\Gamma])}$ is hyperfinite, by arguments similar to that of C2.2 ii), using C3.1.

Consider now the action of Γ on $\mathcal{M}_3([S_\Gamma])$. As in the proof of C2.2, if $\mu \in \mathcal{M}_3([S_\gamma])$, $\mu^3([S_\Gamma]^3) > 0$, and if φ is as in C3.2

$$\nu = \varphi_*((\mu^3|[S_\Gamma]^3)/\mu^3([S_\Gamma]^3))$$

is a measure on $[\Gamma]^{<\infty}$, so let

$$\theta(\mu) = \bigcup\{F \in [\Gamma]^{<\infty} : \nu(\{F\}) \text{ is maximum}\}.$$

Then $\theta(\mu) \in [\Gamma]^{<\infty}$ and θ is a Borel Γ-map from $\mathcal{M}_3([S_\Gamma])$ to $[\Gamma]^{<\infty}$. Thus the equivalence relation $E_\Gamma^{\mathcal{M}_3([S_\Gamma])}$ is tame and the stabilizers of the action of Γ on $\mathcal{M}_3([S_\Gamma])$ are finite.

We now consider the case where there are infinitely many $A_0, A_1, \ldots$ In this case we work with the compact space $[S_\Gamma]^*$ instead of $[S_\Gamma]$. Recall that $[S_\Gamma]^*$ is the disjoint union of the two Γ-invariant Borel sets $[S_\Gamma]$ and S_Γ. Since the action of Γ on $[S_\Gamma]$ and $[S_\Gamma]^2$ has associated equivalence relation hyperfinite, it is easy to verify that the action of Γ on $\mathcal{M}_{\leq 2}([S_\Gamma]^*)$ has also associated equivalence relation hyperfinite. It is also clear from C3.1 that the stabilizers of the action of Γ on $\mathcal{M}_{\leq 2}([S_\Gamma]^*)$ are cyclic-by-finite.

Finally, we consider the action of Γ on $\mathcal{M}_3([S_\Gamma]^*)$. We split $\mathcal{M}_3([S_\Gamma]^*)$ into the two Γ-invariant Borel sets

$$\mathcal{M}' = \{\mu \in \mathcal{M}_3([S_\Gamma]^*) : \mu(S_\Gamma) > 0\}$$

$$\mathcal{M}'' = \{\mu \in \mathcal{M}_3([S_\Gamma]^*) : \mu(S_\Gamma) = 0\}.$$

Then, by arguments similar to the case of $\mathcal{M}_3([S_\Gamma])$ above, we see that there is a Borel Γ-map from $\mathcal{M}'$ to $[\Gamma]^{<\infty}$ and a Borel Γ-map from $\mathcal{M}''$ to $[\Gamma]^{<\infty}$, so there is a Borel Γ-map from $\mathcal{M}_3([S_\Gamma]^*)$ to $[\Gamma]^{<\infty}$. Thus as before, all the stabilizers of the Γ-action on $\mathcal{M}_3([S_\Gamma]^*)$ are finite and $E_\Gamma^{\mathcal{M}_3([S_\Gamma]^*)}$ is tame. ⊣

C4. Hyperbolic groups

For the definition and basic properties of hyperbolic groups, we refer the reader to [Gr], [GdlH], [CPD], [KB]. Free groups are well as finite free products of finite groups are hyperbolic. We will summarize below various facts about hyperbolic groups that we use in this paper, providing appropriate references.

Theorem C4.1 (see [GdlH, 8.37). *Suppose H is a hyperbolic group and $G \leq H$ is a subgroup. Then one of the following holds:*

(i) G is finite,

(ii) G is $\mathbb{Z}$-by-finite (i.e., contains an infinite cyclic subgroup of finite index),

(iii) G contains a copy of F_2.

From this, one has the following corollary, whose proof was supplied to us by Simon Thomas.

Corollary C4.2. *If H is a hyperbolic group which is not amenable, then H does not contain an infinite normal amenable subgroup. In particular if H is torsion-free, it contains no non-trivial normal amenable subgroups.*

Proof. Suppose $N \trianglelefteq H$ is an infinite normal amenable subgroup. Then, by C4.1, N contains an infinite cyclic subgroup C of finite index. We then claim that if $\gamma \in H$, then for some $n \geq 1, \gamma^n \in C$. From this it immediately follows that H does not contain a copy of F_2, so, by C4.1 again, H is itself $\mathbb{Z}$-by-finite, thus amenable, a contradiction.

To prove the claim, assume it fails, and consider the subgroup $\langle\gamma\rangle N$ of H. It is amenable, so it contains an infinite cyclic subgroup $\langle\delta_0\rangle$ of finite index. Clearly $\delta_0 \notin N$, say $\delta_0 = \gamma^{n_0} g_0$, where $n_0 \neq 0, g_0 \in N$. Then $\delta_0^k = \gamma^{n_0 k} g_k$, for some $g_k \in N$. So $\langle\delta_0\rangle \cap N = \{1\}$, and thus $\langle\delta_0\rangle$ has infinite index in N, a contradiction. ⊣

If H is hyperbolic, we denote by ∂H its boundary, a compact, metrizable space. The group H acts continuously on ∂H. We summarize below the relevant for us properties of this action and the induced action on $\mathcal{M}_{\leq 2}(\partial H)$ and $\mathcal{M}_3(\partial H)$.

Theorem C4.3. *Suppose H is hyperbolic.*

i) (See [KB, 4.2] and the proof of C2.2, ii)) If $\gamma \in H$ is torsion-free, then γ fixes exactly two elements of ∂H. In particular, if H is torsion-free, the set of elements of ∂H with non-trivial stabilizers is countable and if we consider the action of H on the set of elements of $\mathcal{M}_{\leq 2}(\partial H)$ with non-trivial stabilizer, the corresponding equivalence relation is tame.

ii) (Adams [Ad94, 5.1], [Ad96, 3.3 and 3.7]) The action of H on ∂H has associated equivalence relation $E_H^{\partial H}$ which is μ-hyperfinite for every measure μ on ∂H, and all the stabilizers are amenable (therefore cyclic-by-finite). Similarly, for the action of H on $\mathcal{M}_{\leq 2}(\partial H)$, the associated equivalence relation $E_H^{\mathcal{M}_{\leq 2}(\partial H)}$ is λ-hyperfinite for every measure λ on $\mathcal{M}_{\leq 2}(\partial H)$ and all the stabilizers are amenable.

iii) (See [Ad96, 5.3]) The action of H on $\mathcal{M}_3(\partial H)$ has finite stabilizers and the corresponding equivalence relation $E_H^{\mathcal{M}_3(\partial H)}$ is tame.

Corollary C4.4. *Every hyperbolic group is near-hyperbolic.*

APPENDIX D

$\mathcal{K}$-structured Equivalence Relations

We use the terminology and notation of [JKL, Section 2.5]. For L a countable relational language and $\mathcal{K}$ a class of countable L-structures, closed under isomorphism, we refer to a countable Borel equivalence relation E on X together with a Borel assignment to each E-class C of a structure $\mathcal{A}_C = \langle C, \dots \rangle \in \mathcal{K}$ as a *$\mathcal{K}$-structured equivalence relation.* We denote this by $(E, \mathcal{A}_C)_{C \in X/E}$. If E admits such $(E, \mathcal{A}_C)_{C \in X/E}$, we call E *$\mathcal{K}$-structurable.*

We now review some notions from Gaboriau [Ga01].

Suppose E is a countable Borel equivalence relation on X. An *E-space* consists of:

i) A standard Borel space U and a surjective Borel map $\pi : U \to X$ with each fiber $\pi^{-1}(\{x\})$ countable. The triple (X, U, π) is called a *standard fiber space* over X.

ii) A Borel map which assigns to each $(x, y) \in E$ and $u \in U$ with $u \in \pi^{-1}(\{y\})$, an element $(x, y) \cdot u \in \pi^{-1}(\{x\})$ such that if $xEyEz$, then for any $u \in \pi^{-1}(\{z\}), (x, y)\cdot((y, z)\cdot u) = (x, z)\cdot u$, and for any $x \in X, u \in \pi^{-1}(\{x\}), (x, x)\cdot u = u$.

Notice that, for each fixed xEy, the map $u \mapsto (x, y)\cdot u$ is a bijection of $\pi^{-1}(\{y\})$ with $\pi^{-1}(\{x\})$ with inverse $v \mapsto (y, x) \cdot v$ and (by definition) $u \mapsto (x, x) \cdot u$ is the identity on $\pi^{-1}(\{x\})$.

Define the following equivalence relation R_E^U on U:

$$uR_E^U v \Leftrightarrow \exists (x, y) \in E[(x, y) \cdot u = v].$$

Thus the R_E^U-class of u is the E-orbit $\{(x, y)\cdot u : (x, y) \in E, \pi(u) = y\}$ of u. Clearly R_E^U is a countable Borel equivalence relation on U. If R_E^U is tame, i.e., admits a Borel transversal, we call the E-space U *tame* (or *discrete* according to Gaboriau).

A *$\mathcal{K}$-structured standard fiber space* consists of a standard fiber space (X, U, π) together with a Borel assignment $x \mapsto \mathcal{A}_x = \langle \pi^{-1}(\{x\}), \dots \rangle \in \mathcal{K}$. A *$\mathcal{K}$-structured E-space* consists of an E-space (X, U, π) such that (X, U, π) is $\mathcal{K}$-structured and for each $(x, y) \in E$ the map $u \mapsto (x, y) \cdot u$ is an isomorphism of $\mathcal{A}_y$ with $\mathcal{A}_x$.

Proposition D.1. *Let E be countable Borel equivalence relation on X. Then the following are equivalent:*

(i) E admits a tame $\mathcal{K}$-structured E-space.

(ii) E can be Borel reduced to a $\mathcal{K}$-structurable countable Borel equivalence relation.

Proof. (i)$\Rightarrow$(ii): Let $(X, U, \pi), (\mathcal{A}_x)_{x \in X}$ be a tame $\mathcal{K}$-structured E-space. Let Y be a Borel transversal for R_E^U.

Define the following equivalence relation F on Y

$$uFv \Leftrightarrow \pi(u)E\pi(v).$$

First we claim that $E \leq_B F$:

Let $\{g_n\}$ be a countable set of Borel functions such that $xEy \Leftrightarrow \exists n(g_n(x) = y)$. Let also $h : X \to U$ be a Borel function with $h(x) \in \pi^{-1}(\{x\})$. Given now $x \in X$, let $n(x)$ be the least n such that $(g_n(x), x)\cdot h(x) \in Y$. Put $f(x) = (g_{n(x)}(x), x)\cdot h(x)$. Then clearly $xEy \Leftrightarrow f(x)Ff(y)$.

So it is enough to show that F is $\mathcal{K}$-structurable. Fix an F-equivalence class $C \subseteq Y$, in order to assign, in a Borel way, a structure $\mathcal{B}_C = \langle C, S^{\mathcal{B}_C}\rangle_{S\in L} \in \mathcal{K}$. Fix $S \in L$, an n-ary relation symbol. All $\pi(u), u \in C$, are in the same E-class D, so choose an element $x_0 \in D$. Given now $u_1, \dots, u_n \in C$, consider $\pi(u_i) = x_i, i = 1, \dots, n$ and $v_i = (x_0, x_i) \cdot u_i \in \pi^{-1}(\{x_0\})$. Put

$$(u_1, \dots, u_n) \in S^{\mathcal{B}_C} \Leftrightarrow (v_1, \dots, v_n) \in S^{\mathcal{A}_{x_0}}.$$

It only remains to show that this definition is independent of the choice of $x_0 \in D$. Suppose another $x'_0 \in D$ was chosen. Let $v'_i = (x'_0, x_i) \cdot u_i \in \pi^{-1}(\{x'_0\})$. We have to show that

$$(v_1, \dots, v_n) \in S^{\mathcal{A}_{x_0}} \Leftrightarrow (v'_1, \dots, v'_n) \in S^{\mathcal{A}_{x'_0}}.$$

For that is enough to check that $(x'_0, x_0) \cdot v_i = v'_i, i = 1, \dots n$. But this is clear as

$$\begin{aligned}(x'_0, x_0) \cdot v_i &= (x'_0, x_0) \cdot (x_0, x_i) \cdot u_i \\ &= (x'_0, x_i) \cdot u_i = v'_i.\end{aligned}$$

Remark. What is really shown here is that (in the obvious sense) U/R^U_E is $\mathcal{K}$-structurable. The tameness of R^U_E is only used to guarantee that this is a standard Borel space.

(ii)⇒(i): Let $(F, \mathcal{B}_C)_{C\in Y/F}$ be a $\mathcal{K}$-structured countable Borel equivalence relation on Y. Let also $f : X \to Y$ be Borel with

$$xEy \Leftrightarrow f(x)Ff(y).$$

Let $U \subseteq X \times Y$ be defined by

$$(x, y) \in U \Leftrightarrow yFf(x).$$

Let for $(x, y) \in U, \pi(x, y) = x$. Clearly (X, U, π) is a standard fiber space. Given $x \in X$, define

$$\begin{aligned}\mathcal{A}_x &= \langle \pi^{-1}(\{x\}), S^{\mathcal{A}_x}\rangle_{S\in L} \\ &= \langle \{x\} \times [f(x)]_F, S^{\mathcal{A}_x}\rangle_{S\in L}\end{aligned}$$

as follows: Given $(x, y_1), \dots, (x, y_n) \in \pi^{-1}(\{x\})$, let

$$S^{\mathcal{A}_x}((x, y_1), \dots, (x, y_n)) \Leftrightarrow (y_1, \dots, y_n) \in S^{\mathcal{B}_C},$$

where $C = [f(x)]_F$. Thus

$$(x, y) \mapsto y$$

is an isomorphism between $\mathcal{A}_x$ and $\mathcal{B}_{[f(x)]_F}$.

We finally define the E-action on this fiber space:

Given $(x_1, x_2) \in E$ and $(x_2, y) \in \pi^{-1}(\{x_2\})$, let

$$(x_1, x_2) \cdot (x_2, y) = (x_1, y).$$

Clearly $u \mapsto (x_1, x_2) \cdot u$ is an isomorphism from $\mathcal{A}_{x_2}$ to $\mathcal{A}_{x_1}$, so the only thing we need to check is that this E-space is tame. But clearly for $(x_1, y_1), (x_2, y_2) \in U$,

$$(x_1, y_1)R^U_E(x_2, y_2) \Leftrightarrow y_1 = y_2.$$

(Notice that $(x_1, y_1), (x_2, y_2) \in U, y_1 = y_2$, imply that $f(x_1) F f(x_2)$, thus $(x_1, x_2) \in E$). So R^U_E is indeed tame. ⊣

We note next the following fact.

Proposition D.2. *Suppose $E \subseteq F$ are countable Borel equivalence relations. If F admits a tame $\mathcal{K}$-structured space, so does E.*

Corollary D.3. *If $E \subseteq F$ are countable Borel equivalence relations and F can be Borel reduced to a $\mathcal{K}$-structurable countable Borel equivalence relation, so can E.*

Proof of Proposition D.2. Let $(X, U, \pi), (\mathcal{A}_x)_{x \in X}$ be a smooth $\mathcal{K}$-structured F-space. The restriction of the F-action to E clearly gives a $\mathcal{K}$-structured E-space and it only remains to show that it is tame, i.e., R^U_E is tame. But note that $R^U_E \subseteq R^U_F$ and R^U_F is a countable tame equivalence relation, thus so is R^U_E. ⊣

Of particular interest to us in this paper is the class $\mathcal{K}_m$ of m-dimensional contractible (abstract) simplicial complexes (see, e.g., [L, p. 96]). A simplicial complex consists of a non-empty countable set A and for each $k = 1, 2, \ldots,$ a collection S_k of nonempty subsets of A of cardinality $k+1$, such that letting $S_0 = A$, we have that every k element subset of any $B \in S_k$ is in S_{k-1}. The elements of S_k are called the *k-simplexes* of the simplicial complex. If there is a largest m such that $S_m \neq \emptyset$, we say that the simplicial complex is *m-dimensional.* An m-dimensional complex $\langle A, S_k \rangle_{k=1,2,\ldots,m}$ can be represented as a structure $\mathcal{A}$ in the countable language $L_m = \{R_1, \ldots, R_m\}$ where R_i is an $(i+1)$-ary relation symbol, by letting

$$\mathcal{A} = \langle A, R^{\mathcal{A}} \rangle_{i=1,\ldots,m},$$

where

$$R^{\mathcal{A}}_i(a_0, \ldots, a_i) \Leftrightarrow \{a_0, \ldots, a_i\} \in S_i.$$

We say that a simplicial complex is contractible if its geometric realization, (see, e.g., [L, p. 98]) is contractible. We denote by $\mathcal{K}_m$ the class of m-dimensional contractible simplicial complexes. (For $m = 0$, $\mathcal{K}_0$ consists of the trivial structures with one element and no relations.) A one-dimensional contractible simplicial complex is simply a tree (see, e.g., [Ma, Section 6.4], thus $\mathcal{K}_1$ can be identified with the class $\mathcal{T}$ of trees.

Using the concept of a product of simplicial complexes (see, e.g., [EDM2, p. 262]), one can see that if $E_1, \ldots, E_m$ are equivalence relations, where each E_i is induced by a free Borel action of F_2, then $E_1 \times \cdots \times E_m$ is $\mathcal{K}_m$-structurable.

APPENDIX E

Proof of the General Case of Theorem 4.4

E1. Amenable classes of structures

In what follows we will adopt the convention that whenever we consider a class of structures $\mathcal{C}$ we have in mind that they are all on the fixed set $\mathbb{N}$.

We say that a Borel class of structures $\mathcal{C}$, closed under isomorphism, is *essentially countable* if

$$(\cong|\mathcal{C}) \leq_B E_\infty;$$

that is to say, if its isomorphism relation is Borel reducible to the universal countable Borel equivalence relation of [DJK], or equivalently, there is some countable Borel equivalence relation to which it is reducible. We say that a class of countable structures $\mathcal{C}$ is *measure amenable* if there is an assignment

$$\mathcal{M} \mapsto \varphi_{\mathcal{M}}$$

of means on $\mathbb{N}$ to structures in $\mathcal{C}$ such that whenever $\pi : \mathcal{M} \cong \mathcal{N}$ is an isomorphism, then

$$\varphi_{\mathcal{N}} \circ \pi = \varphi_{\mathcal{M}},$$

and moreover the map $\Phi : \mathcal{C} \times [-1,1]^{\mathbb{N}} \to [-1,1]$ given by

$$(\mathcal{M}, f) \mapsto \varphi_{\mathcal{M}}(f)$$

is universally measurable.

For example: Assume the Continuum Hypothesis, and let $\mathcal{C}$ be the structures $\mathcal{M}$ on $\mathbb{N}$ which are isomorphic to

$$(\mathbb{Z}, <, P),$$

the integers under the usual ordering, equipped with some unary predicate $P \subseteq \mathbb{Z}$ which will depend on the particular $\mathcal{M} \in \mathcal{C}$. For each $\mathcal{M} \in \mathcal{C}$, we can, in a Borel manner, choose an isomorphism

$$\psi : \mathcal{M} \cong \mathbb{Z},$$

and then, for each n, let

$$\varphi_{\mathcal{M},n} : [-1,1]^{\mathcal{M}} \to [-1,1]$$

$$f \mapsto \frac{1}{2n+1} \sum_{i=-n}^{i=n} f(\psi^{-1}(i)).$$

Using Christensen, Mokobodzki, as in 0F, we may find a universally measurable shift-invariant mean φ on $\mathbb{N}$ and for each $\mathcal{M}$ let

$$\varphi_{\mathcal{M}}(f) = \varphi(n \mapsto \varphi_{\mathcal{M},n}(f)).$$

We have to face up to the fact that $\mathcal{C}$ may be measure amenable without being essentially countable: For instance, consider the class $\mathcal{C}$ of the models of [Mak], consisting of isomorphic copies of expansions of $(\mathbb{N}\cdot\mathbb{Z}, <^{\mathrm{lex}})$; as discussed in [Hj00a], this isomorphism relation is far from essentially countable, but if we assume CH then we can, in parallel to the example above, show that it is measure amenable.

The counterexamples can get worse. The class $\mathcal{C}$ of countable structures may be amenable and essentially countable, without $\cong|\mathcal{C}$ being Borel reducible or universally measurably reducible to an amenable countable Borel equivalence relation. If we take any class of structures $\mathcal{B}$ whatsoever and replace it by $\mathcal{C} = \{\mathcal{M} \sqcup \langle\mathbb{Z}, <\rangle : \mathcal{M} \in \mathcal{B}\}$ (appropriately coded as structures on $\mathbb{N}$) then, under CH, we may assign suitably invariant means by concentrating on the $\langle\mathbb{Z}, <\rangle$, while maintaining $(\cong|\mathcal{B}) \leq_B (\cong|\mathcal{C})$. This difficulty is the subject of the lemmas below. Under suitable assumptions, we do indeed obtain that essential countability will entail reduction to an amenable countable Borel equivalence relation.

The second problem is that without CH there is no known way to obtain any examples of measure amenable structures. This we will bypass by the now standard metamathematical trick. As mentioned in the course of 4.4, the statements being proved are all projective, and this licenses the use of CH.

The next lemma is implicit in [HK].

Lemma E1.1. *Let $\mathcal{C}$ be a class of structures with*

$$\theta : \mathcal{C} \to X_\infty$$

witnessing that $(\cong|\mathcal{C}) \leq_B E_\infty$. Then we may find a countable Borel equivalence relation F on a standard Borel Y with $\gamma : \mathcal{C} \to Y$ witnessing $(\cong|\mathcal{C}) \leq_B F$, and we may, in a Borel manner, assign to each $\mathcal{M} \in \mathcal{C}$ a pair $(A_\mathcal{M}, \rho_m)$ such that

(a) $A_\mathcal{M} \subseteq \bigcup_n \mathcal{M}^n$;

(b) $\rho_\mathcal{M} : A_\mathcal{M} \to Y$;

(c) $\mathcal{M} \mapsto (A_\mathcal{M}, \rho_\mathcal{M})$ is $\cong$-invariant, in the sense that if $\pi : \mathcal{M} \cong \mathcal{N}$ then

$$\pi(A_\mathcal{M}) = A_\mathcal{N},$$

$$\rho_\mathcal{M} = \rho_\mathcal{N} \circ \pi;$$

(d) $\rho_\mathcal{M}(\vec{a}) F \gamma(\mathcal{M})$, for all $\vec{a} \in A_\mathcal{M}$;

(e) Y equals the range of γ;

(f) there is Borel $\psi : Y \to \mathcal{C}$ such that $\psi(\gamma(\mathcal{M})) \cong \mathcal{M}$, for all $\mathcal{M} \in \mathcal{C}$.

Proof. Following [HK], we may find a countable fragment $\mathcal{F} \subseteq \mathcal{L}_{\omega_1,\omega}$ such that for each $\mathcal{M} \in \mathcal{C}$ there will be some $\vec{a} \in \mathcal{M}^{<\mathbb{N}}$ such that $\langle\mathcal{M}, \vec{a}\rangle$ is $\mathcal{F}$-atomic. The following are all routine consequences of the definition of atomicity:

(i) the set $\{\langle\mathcal{M}, \vec{a}\rangle : \langle\mathcal{M}, \vec{a}\rangle$ is $\mathcal{F}$-atomic $\}$ is Borel;

(ii) in a given fragment $\mathcal{F}_n = \mathcal{F}(c_1, ..., c_n)$ (where $c_1, ..., c_n$ are fresh constant symbols), the collection of $T \subseteq \mathcal{F}_n$ which are complete and admit an atomic model is Borel;

(iii) for each T as in (ii) we may in a Borel way choose some $\langle\mathcal{M}, \vec{a}\rangle$ with $\langle\mathcal{M}, \vec{a}\rangle \models T$.

Given these facts, we let

$$A_\mathcal{M} = \{\vec{a} : \langle\mathcal{M}, \vec{a}\rangle \text{ is } \mathcal{F}\text{−atomic}\},$$

$$\rho_\mathcal{M}(\vec{a}) = \mathrm{Th}_\mathcal{F}(\langle\mathcal{M}, \vec{a}\rangle),$$

$$Y = \{\mathrm{Th}_F(\langle\mathcal{M}, \vec{a}\rangle) : \vec{a} \in A_\mathcal{M}, \mathcal{M} \in \mathcal{C}\};$$

for each $T \in Y$ we choose in a Borel manner $\mathcal{M} \in \mathcal{C}$ using (iii) and let this be $\psi(T)$; we set $T_1 F T_2$ if $\psi(T_1) \cong \psi(T_2)$; and we let $\gamma(\mathcal{M}) = \rho_{\mathcal{M}}(\vec{a})$, for $\vec{a}$ the first tuple, in the lexicographic ordering on $\mathbb{N}^{<\mathbb{N}}$, with $\vec{a} \in A_{\mathcal{M}}$. ⊣

Lemma E1.2. *Let $\mathcal{C}$ be a measure amenable class of countable structures, and suppose that for all $\mathcal{M} \in \mathcal{C}$ and $a \in \mathcal{M}$,*

$$\mathcal{M} \models \forall b \bigvee_{t \text{ a term}} t(a) = b;$$

that is to say, the algebraic closure of any $a \in \mathcal{M}$ is the whole structure.

Then $\cong |\mathcal{C}$ is Borel bireducible to a measure amenable countable Borel equivalence relation.

Proof. It is easily seen, and observed in the course of [HK], that, since structures in $\mathcal{C}$ are finitely generated, $\cong |\mathcal{C}$ is essentially countable. Thus E1.1 applies and, using its notation, we will show that F is measure amenable.

Let $(t_i)_{i \in \mathbb{N}}$ be some fixed enumeration of terms. Define for each $\mathcal{M} \in \mathcal{C}$

$$\sigma_{\mathcal{M}} : \mathcal{M} \to A_{\mathcal{M}},$$

by

$$\sigma_{\mathcal{M}}(a) = (t_{i(1)}(a), t_{i(2)}(a), ..., t_{i(n)}(a)),$$

where the sequence $i(1), i(2), ..., i(n)$ is chosen of least length and then lexicographically least so that

$$(t_{i(1)}(a), t_{i(2)}(a), ..., t_{i(n)}(a)) \in A_{\mathcal{M}}.$$

Clearly, $\mathcal{M} \mapsto (A_{\mathcal{M}}, \sigma_{\mathcal{M}})$ is isomorphism invariant.

To see that F is measure amenable, notice that each F-class D is of the form $\{\mathrm{Th}_{\mathcal{F}}(\langle \mathcal{M}, \vec{a} \rangle) : \vec{a} \in A_{\mathcal{M}}\}$, for some $\mathcal{M} \in \mathcal{C}$, and that if it is also equal to $\{\mathrm{Th}_{\mathcal{F}}(\langle \mathcal{N}, \vec{b} \rangle) : \vec{b} \in A_{\mathcal{N}}\}$ for some $\mathcal{N} \in \mathcal{C}$, then $\mathcal{M} \cong \mathcal{N}$.

We define a mean ψ_D on D as follows. Fix a universally measurable assignment $\mathcal{M} \mapsto \varphi_{\mathcal{M}}$ that verifies that $\mathcal{C}$ is measure amenable. Then given $f \in \ell_\infty(D)$, put

$$\psi_D(f) = \varphi_{\mathcal{M}}(a \in \mathcal{M} \mapsto f(\mathrm{Th}_{\mathcal{F}}(\langle \mathcal{M}, \vec{a} \rangle))).$$

Using the invariance properties of $\varphi_{\mathcal{M}}$, it is easy to see that ψ_D is well-defined, i.e., independent of the choice of $\mathcal{M}$. The verification of universal measurability is routine. ⊣

Remark. A similar argument works if we replace the hypothesis that the algebraic closure of each $a \in \mathcal{M}$ is the whole structure, by the hypothesis that, for some n, the algebraic closure of each $\vec{a} \in \mathcal{M}^n$ is the whole structure.

E2. The factoring lemma

We prove here the general case of 4.4.

Lemma E2.1 *Let Γ and $H = H_1 \times H_2 \times \dots H_n$ be countable groups acting by Borel transformations on standard Borel spaces X and Y, resp., with the H acting freely, and let*

$$\rho : X \to Y$$

be a Borel homomorphism of E_Γ^X to E_H^Y with

$$\alpha : X \times \Gamma \to H$$

the associated cocycle, and

$$\alpha_i = p_i \circ \alpha : X \times \Gamma \to H_i$$

the induced cocycle for the various H_i.

Suppose that for each $i \leq n$ *we have either:*

(i) α_i *maps into an amenable subgroup of* H_i, *or*

(ii) there is a hyperfinite equivalence relation E_i *on some standard Borel* Z_i *and a Borel homomorphism* $\rho_i : X \to Z_i$ *from* E_Γ^X *to* E_i *with the property that for all* $x \in X, \gamma \in \Gamma$

$$\rho_i(\gamma \cdot x) = \rho_i(x) \Leftrightarrow \alpha_i(\gamma, x) = 1.$$

Then there is a countable Borel equivalence relation $\hat{F}$ *on* $\hat{Z}$, *such that* ρ *factors through* $\hat{F}$ *and* $\hat{F}$ *is* $\hat{\mu}$*-hyperfinite for any measure* $\hat{\mu}$ *on* $\hat{Z}$.

Proof. We may as well assume that $n = 2$, H_1 is amenable, and that there is a Borel homomorphism

$$(\rho_2 =)\rho' : X \to Z(= Z_2)$$

from E_Γ^X to some hyperfinite $F(= E_2)$ with the additional property that for all $x \in X, \gamma \in \Gamma$

$$\rho'(\gamma \cdot x) = \rho'(x) \Leftrightarrow \alpha_2(\gamma, x) = 1.$$

We make the further harmless assumption that each $\rho'([x]_\Gamma)$ is infinite; the case for those which are finite can be dealt with separately using an argument which is similar to and somewhat easier than the one below.

Since hyperfinite equivalence relations are induced by $\mathbb{Z}$-actions (see [DJK], [JKL]), we may in a Borel fashion assign

$$z \mapsto <_z,$$

where each $<_z$ is a linear order of order type $\mathbb{Z}$ on $[z]_F$ with the invariance property that

$$z_1 F z_2 \Rightarrow <_{z_1} = <_{z_2}.$$

We then write

$$x R_\ell x'$$

if $\rho'(x) F \rho'(x')$ and $\rho'(x)$, $\rho'(x')$ are exactly ℓ many places apart in the linear order $<_{\rho'(x)} = <_{\rho'(x')}$ restricted to $\rho'([x]_\Gamma)$. In other words, if there are $x_0, x_1, \ldots, x_\ell \in [x]_\Gamma$ with

$$x_0 = x,$$
$$x_\ell = x',$$
$$\rho'(x_0) = \rho'(x),$$
$$\rho'(x_\ell) = \rho'(x'),$$

each

$$\rho'(x_i) <_{\rho'(x)} \rho'(x_{i+1})$$

and for all $u \in [x]_\Gamma$ with

$$\rho'(x_0) <_{\rho'(x)} \rho'(u) <_{\rho'(x)} \rho'(x_\ell)$$

there will be some i with

$$\rho'(\hat{x}) = \rho'(x_i).$$

This definition makes sense for $\ell \geq 0$, and we can extend it in the natural way to $\ell < 0$ by setting

$$x' R_\ell x$$

if $xR_{(-\ell)}x'$.

We will build a measure amenable class of structures $\mathcal{C}$ and then a class of expansions $\mathcal{D}$. We will show that ρ factors through $\cong|\mathcal{D}$ and then appeal to E1.2 to argue that $\cong|\mathcal{D}$ is Borel bireducible to a measure amenable countable Borel equivalence relation.

The language for $\mathcal{C}$ will have unary functions $(F_\ell)_{\ell\in\mathbb{Z}}$, $(\hat{F}_h)_{h\in H_1}$, and an equivalence relation E^*. The structures in $\mathcal{C}$ are those structures on $\mathbb{N}$ that satisfy:

(i) $\forall a([a]_{E^*} = \{\hat{F}_h(a) : h \in H_1\})$; for all $h_1, h_2 \in H_1$,

$$\hat{F}_{h_1} \circ \hat{F}_{h_2} = \hat{F}_{h_1+h_2},$$

$$\forall b(\hat{F}_0(b) = b);$$

(ii) $\forall a, b$,

$$\bigvee_{\ell\in\mathbb{Z}} \bigvee_{h\in H_1} (\hat{F}_h \circ F_\ell(a) = b);$$

(iii) for all $\ell \in \mathbb{Z}, h_1, h_2, h_3, h_4 \in H_1$,

$$\forall a, b \bigvee_{\bar{h}\in H_1} (\hat{F}_{\bar{h}} \circ \hat{F}_{h_1} \circ F_\ell \circ \hat{F}_{h_2}(a) = \hat{F}_{h_3} \circ F_\ell \circ \hat{F}_{h_4}(a));$$

(iv)

$$\forall a \bigvee_{h\in H_1} (\hat{F}_h(a) = F_0(a));$$

(v) for all $\ell_1, \ell_2 \in \mathbb{Z}$,

$$\forall a \bigvee_{h\in H_1} (\hat{F}_h \circ F_{\ell_1+\ell_2}(a) = F_{\ell_1} \circ F_{\ell_2}(a)).$$

In essence, (i) states that the $\{\hat{F}_h : h \in H_1\}$ act transitively on each E^*-equivalence class, and that the action respects the group structure suggested by H_1. (ii) states in particular that the algebraic closure of any point is the whole structure. (iii)-(v) state that in essence that $\mathbb{Z}$, via the assignment $\ell \mapsto F_\ell$, acts on the structure's collection of all E^* equivalence classes.

We will in fact see that this class of structures is not only measure amenable but "2-amenable" in the sense of [JKL].

We begin with a Følner sequence $(D_n)_{n\in\mathbb{N}}$ for H_1. For $n \in \mathbb{N}$, $\mathcal{M} \in \mathcal{C}$, $a \in \mathcal{M}$, we first let

$$f_n^{a,\mathcal{M}} = \frac{1}{|D_n|} 1_{\{F_h(a):h\in D_n\}};$$

that is to say, $f_n^{a,\mathcal{M}}(b)$ equals $1/D_n$ if there exists $h \in D_n$ with $\mathcal{M} \models \hat{F}_h(a) = b$, and equals zero otherwise. We iterate and define for each n, m,

$$f_{n,m}^{a,\mathcal{M}} : \mathcal{M} \to [0,1]$$

$$f_{n,m}^{a,\mathcal{M}} = \frac{1}{2m+1}(\sum_{\ell\in[-m,m]} f_n^{F_\ell(a),\mathcal{M}}).$$

For each $\mathcal{M} \in \mathcal{C}$ and $a, b \in \mathcal{M}$ with aE^*b, it follows from (i) and the properties of the Følner sequence that

$$\lim_{n\to\infty} \| f_n^{a,\mathcal{M}} - f_n^{b,\mathcal{M}} \|_{\ell_1} \to 0,$$

and then by (ii) we have that for *all* $a, b \in \mathcal{M}$,

$$\lim_{m\to\infty}(\lim_{n\to\infty} \| f_{n,m}^{a,\mathcal{M}} - f_{n,m}^{b,\mathcal{M}} \|_{\ell_1}) \to 0.$$

We massage this into the form required by [JKL, 2.5] by recalling that each $\mathcal{M} \in \mathcal{C}$ has $\mathbb{N} = \{0, 1, 2, 3, \dots\}$ as its underlying set, and letting

$$f^{\mathcal{M}}_{n,m} = f^{0,\mathcal{M}}_{n,m},$$

and thereby obtain that $\mathcal{C}$ is 2-amenable. It then follows that we can integrate against an invariant universally measurable mean to obtain that $\mathcal{C}$ is measure amenable (see [Ke91], [JKL], or the example at the start of E.1).

The class $\mathcal{D}$ arises by expanding the structures in $\mathcal{C}$ through the addition of fresh unary predicates, $(U_n)_{n\in\mathbb{N}}$. For this purpose let us fix a countable Borel separating family $(B_n)_{n\in\mathbb{N}}$ for Y. To each $\mathcal{M} \in \mathcal{C}$ and function

$$\sigma : \mathcal{M} \to Y$$

we associate the structure $\mathcal{M}^\sigma$ which includes the language of $\mathcal{C}$, on which it behaves just like $\mathcal{M}$, but interprets the fresh unary predicates by obeying the dictate that

$$(\mathcal{N}^\sigma \models U_n(a)) \Leftrightarrow \sigma(a) \in B_n.$$

We let $\mathcal{D}$ be the collection $\{\mathcal{M}^\sigma | \mathcal{M} \in \mathcal{C}, \sigma : \mathcal{M} \to Y\}$. These satisfy the assumptions of E1.2 and so it suffices to show that ρ factors through $\cong |\mathcal{D}$.

For each x, we define a pair $n_x, \sigma_x : \mathcal{N}_x \to Y$.

The underlying set of $\mathcal{N}_x$ will be the set of pairs (ℓ, y) such for some $u \in [x]_\Gamma$ we have

(a) $xR_\ell u$;

(b) $yE_{H_1}\rho(u)$.

We set

$$(\ell, y)E^*(\bar{\ell}, \bar{y})$$

if and only if $\ell = \bar{\ell}$. We naturally enough define the $\hat{F}_h$ functions as suggested by the action of H_1, so that

$$\hat{F}_h(\ell, y) = (\ell, h \cdot y).$$

The F_ℓ functions require more care.

First fix an enumeration $(\bar{h}_i)_{i\in\mathbb{N}}$ of $H_1 \times H_2$. Then let

$$F_\ell((y, \bar{\ell})) = (\bar{h}_{i_0} \cdot y, \ell + \bar{\ell}),$$

where i_0 is least such that for some $u \in [x]_\Gamma$ we have

$$xR_{\ell+\bar{\ell}}u,$$

$$(\bar{h}_{i_0} \cdot y)E_{H_1}\rho(u).$$

We then, in the most obvious way possible, define

$$\sigma_x : \mathcal{N}_x \to Y$$

$$(\ell, y) \mapsto y.$$

In a Borel manner it is possible to choose for each x a bijection

$$\pi_x : \mathbb{N} \cong \{(\ell, y) : \ell \in \mathbb{Z}, \exists u \in [x]_\Gamma(yE_{H_1}\rho(u), xR_\ell u)\},$$

and we finish the definition by letting $\mathcal{N}^*_x$ be the copy on $\mathbb{N}$ of $\mathcal{N}_x$ provided by π_x and then setting

$$\mathcal{M}_x = (\mathcal{N}^*_x)^{\sigma_x \circ \pi_x}.$$

Claim (I): If $u_1, u_2 \in [x]_\Gamma$ with

$$xR_\ell u_1,$$

$$xR_\ell u_2,$$

then

$$[\rho(u_1)]_{H_1} = [\rho(u_2)]_{H_1}.$$

Proof of claim: Fix $\gamma \in \Gamma$ with $\gamma \cdot u_1 = u_2$. It follows from the definition of R_ℓ that $\rho'(u_1) = \rho'(u_2)$, and hence

$$\alpha_2(\gamma, u_1) = 1$$
$$\therefore \alpha(\gamma, u_1) \in H_1 \times \{1\}$$
$$\therefore [\rho(u_1)]_{H_1} = [\rho(u_2)]_{H_2}.$$

(⊣Claim)

Claim (II): For any $x \in X$ we have $\mathcal{N}_x^*$ in $\mathcal{C}$.

Proof of claim: Appealing to Claim (I), we have

$$\{y : (\ell, y) \in \mathcal{N}_x\}$$

is an H_1-equivalence class for any $\ell \in Z$. This quickly implies (i) in the definition of $\mathcal{C}$. (ii) follows from the careful choice of the F_ℓ functions. For (iii)-(v) we observe that the E^*-equivalence classes are arranged into a $\mathbb{Z}$-chain by $\{xR_\ell(\cdot) : \ell \in \mathbb{Z}\}$ and that the action of the F_ℓ functions respects that ordering. (⊣Claim)

Claim (III): $\mathcal{M}_{x_1} \cong \mathcal{M}_{x_2} \Rightarrow \rho(x_1) E_H \rho(x_2)$.

Proof of claim: Since the functions $\sigma_{x_1} \circ \pi_{x_1}$ and $\sigma_{x_2} \circ \pi_{x_2}$ must have a point in common. (⊣Claim)

Claim (IV): $x_1 E_\Gamma^X x_2 \Rightarrow \mathcal{M}_{x_1} \cong \mathcal{M}_{x_2}$.

Proof of claim: If $x_1 E_\Gamma^X x_2$, then we can find some $\hat{\ell}$ with

$$x_1 R_{\hat{\ell}} x_2,$$

and hence for all ℓ

$$x_1 R_{\ell+\hat{\ell}}(\cdot) = x_2 R_\ell(\cdot).$$

We simply define an isomorphism

$$\psi : \mathcal{N}_{x_1} \cong \mathcal{N}_{x_2}$$

by

$$\psi(\ell, y) = (\ell - \hat{\ell}, y).$$

This isomorphism clearly intertwines σ_{x_1} and σ_{x_2}. (⊣Claim)

Thus ρ factors through $\cong |\mathcal{D}$. By E1.2, we have that $\cong |\mathcal{D}$ is Borel bireducible to a measure amenable countable Borel equivalence relation, which by [CFW] will be $\hat{\mu}$-hyperfinite for all measures $\hat{\mu}$. ⊣

Bibliography

[Ad88] S. Adams, Indecomposability of treed equivalence relations, *Israel J. Math.*, **64(3)**, 362–380, 1988.

[Ad94] S. Adams, Boundary amenability for word hyperbolic groups and an application to smooth dynamics of simple groups, *Topology*, **33(4)**, 765–783, 1994.

[Ad94a] S. Adams, Indecomposability of equivalence relations generated by word hyperbolic groups, *Topology*, **33**, 785–798, 1994.

[Ad95] S. Adams, Some new rigidity results for stable orbit equivalence, *Ergod. Th. and Dynam. Syst.*, **15**, 209–219, 1995.

[Ad96] S. Adams, Reduction of cocycles with hyperbolic targets, *Erg. Th. and Dynam. Sys.*, **16**, 1111–1145, 1996.

[Ad02] S. Adams, Containment does not imply Borel reducibility, *Set Theory*, S. Thomas (Ed.), DIMACS Series, **58**, Amer. Math. Soc., 1–23, 2002.

[AK] S. Adams and A.S. Kechris, Linear algebraic groups and countable Borel equivalence relations, *J. Amer. Math. Soc.*, **13(4)**, 909–943, 2000.

[AL] S. Adams and R. Lyons, Amenability, Kazhdan's property and percolation for trees, groups, and equivalence relations, *Israel J. Math.*, **75**, 341–370, 1991.

[BK] H. Becker and A.S. Kechris, *The Descriptive Set Theory of Polish Group Actions*, London Math. Society Lecture Note Series, **232**, Cambridge Univ. Press, 1996.

[BM] M.B. Bekka and M. Mayer, *Ergodic Theory and Topological Dynamics of Group Actions on Homogeneous Spaces*, London Math. Soc. Lecture Note Series, **269**, Cambridge Univ. Press, 2000.

[CCJJV] P.-A. Cherix, M. Cowling, P. Jolissaint, P. Julg, and A. Valette, *Groups with the Haagerup Property*, Progress in Mathematics, **197**, Birkhäuser, 2001.

[C] J.P.R. Christensen, *Topology and Borel Structure*, North-Holland Math. Studies, **10**, North-Holland, 1974.

[CFW] A. Connes, J. Feldman, and B. Weiss, An amenable equivalence relation is generated by a single transformation, *Erg. Th. and Dynam. Syst.*, **1**, 431–450, 1981.

[CW] A. Connes and B. Weiss, Property T and asymptotically invariant sequences, *Israel J. Math.*, **37**, 209–210, 1981.

[CDP] M. Coornaert, T. Delzant, and A. Papadopoulos, *Géométrie et théorie des groupes*, Lecture Notes in Math., **144**, Springer-Verlag, 1990.

[dlH] P. de la Harpe, *Topics in Geometric Group Theory*, Chicago Lectures in Math., Univ. of Chicago Press, 2000.

[dJR] A. del Junco and J. Rosenblatt, Counterexamples in ergodic theory and number theory, *Math. Ann.*, **245**, 185–197, 1979.

[DJK] R. Dougherty, S. Jackson, and A.S. Kechris, The structure of hyperfinite Borel equivalence relations, *Trans. Amer. Math. Soc.*, **341**, 193–225, 1994.

[D] H. Dye, On groups of measure preserving transformations, I, II, *Amer. J. Math.*, **81**, 119–159 and **85**, 551–576, 1963.

[EDM2] *Encyclopedic Dictionary of Mathematics*, 2nd Ed., K. Itô (Ed.), MIT Press, 1996.

[F] R.H. Farrell, Representation of invariant measures, *Ill. J. Math.*, **6**, 447–467, 1962.

[FM] J. Feldman and C.C. Moore, Ergodic equivalence relations and von Neumann algebras, I, *Trans. Amer. Math. Soc.*, **234**, 289–324, 1977.

[Fu] A. Furman, Orbit equivalence rigidity, *Ann. of Math.*, **150**, 1083–1108, 1999.

[Fur] H. Furstenberg, Boundary theory and stochastic processes in homogeneous spaces, *Proc. Sympos. Pure Math.*, **26**, 193–229, 1973.

[Ga00] D. Gaboriau, Coût des relations d'equivalence et des groupes, *Invent. Math.*, **739**, 41–98, 2000.

[Ga01] D. Gaboriau, Invariants ℓ^2 de relations d'equivalence et de groupes, preprint, 2001.

[GdlH] E. Ghys and P. de la Harpe, *Sur les Groupes Hyperboliques d'après Mikael Gromov*, Progress in Math., **83**, Birkhäuser, 1990.

[Gr] M. Gromov, Hyperbolic groups, *Essays in Group Theory*, 75–263, Springer, 1987.

[Hj00a] G. Hjorth, *Classification and Orbit Equivalence Relations*, Math. Surveys and Monographs, **75**, Amer. Math. Soc., 2000.

[Hj00b] G. Hjorth, Wildness in the product groups, *Fund. Math.*, **164**, 1–33, 2000.

[HK] G. Hjorth and A.S. Kechris, Borel equivalence relations and classification of countable models, *Ann. Pure Appl. Logic*, **82**, 221–272, 1996.

[JKL] S. Jackson, A.S. Kechris, and A. Louveau, Countable Borel equivalence relations, *J. Math Logic*, **2(1)**, 1–80, 2002.

[JS] V. Jones and K. Schmidt, Asymptotically invariant sequences and approximate finiteness, *Amer. J. Math.*, **109**, 91–114, 1987.

[KB] I. Kapovich and N. Benakli, Boundaries of hyperbolic groups, preprint, 2001.

[Ka] Y. Katznelson, *An Introduction to Harmonic Analysis*, Dover, 1976.

[Ke91] A.S. Kechris, Amenable equivalence relations and Turing degrees, *J. Symb. Logic*, **56**, 182–194, 1991.

[Ke95] A.S. Kechris, *Classical Descriptive Set Theory*, Springer-Verlag, 1995.

[Ke99] A.S. Kechris, New directions in descriptive set theory, *Bull. Symb. Logic*, **5(2)**, 161–174, 1999.

[Ke01] A.S. Kechris, Lectures on costs of equivalence relations and groups, preprint, 2001.

[L] J.M. Lee, *Introduction to Topological Manifolds*, Springer, 2000.

[Lev] G. Levitt, On the cost of generating an equivalence relation, *Erg. Th. and Dyn. Syst.*, **15**, 1173–1181, 1995.

[LR] V. Losert and H. Rindler, Almost invariant sets, *Bull. London Math. Soc.*, **13**, 145–148, 1981.

[LS] R.C. Lyndon and P.E. Schupp, *Combinational Group Theory*, Springer 2001.

[M] W.S. Massey, *Algebraic Topology: An Introduction*, Springer, 1967.

[Mak] M. Makkai, An example concerning Scott heights, *J. Symb. Logic*, **46**, 301–318, 1981.

[Ma77] G.A. Margulis, Discrete groups of motions of manifolds of non-positive curvature, *Amer. Math. Soc. Transl.*, **109**, 33-45, 1977.

[Ma77a] G.A. Margulis, Arithmeticity of irreducible lattices in semisimple groups of rank greater than 1, Appendix to Russian translation of M. Ragunathan, Discrete subgroups of Lie groups, Mir, Moscow (1977), in Russian.

[Mo] G.D. Mostow, *Strong rigidity of locally symmetric spaces*, Ann. of Math. Studies, **78**, Princeton Univ. Press, 1973.

[MS02] N. Monod and Y. Shalom, *Cocycle superrigidity and bounded cohomology for negatively curved spaces*, preprint, 2002.

[MS02a] N. Monod and Y. Shalom, *Orbit equivalence rigidity and bounded cohomology*, preprint, 2002.

[OW] D. Ornstein and B. Weiss, Ergodic theory of amenable group actions, I, The Rohlin Lemma, *Bull. Amer. Math. Soc.*, **2**, 161–164, 1980.

[P] K.R. Parthasarathy, *Probability Measures on Metric Spaces*, Academic Press, 1967.

[Po] S. Popa, private communication, 2001.

[Ra] M. Ratner, On Raghunathan's measure conjecture, *Ann. of Math.*, **134**, 545–607, 1991.

[R] J. Rosenblatt, Uniqueness of invariant means for measure-preserving transformations, *Trans. Amer. Math. Soc.*, **265**, 623–636, 1981.

[Ro] J.J. Rotman, *An Introduction to the Theory of Groups*, Springer-Verlag, 1995.

[S80] K. Schmidt, Asymptotically invariant sequences and an action of $SL(2, \mathbb{Z})$ on the 2-sphere, *Israel J. Math.*, **37**, 193–207, 1980.

[S81] K. Schmidt, Amenability, Kazhdan's Property T, strong ergodicity and invariant means for ergodic group actions, *Erg. Th. and Dyn. Systems*, **1**, 223–236, 1981.

[S84] K. Schmidt, Asymptotic properties of unitary representations and mixing, *Proc. London Math. Soc.*, **48**, 445–460, 1984.

[SW] K. Schmidt and P. Walters, Mildly mixing actions of locally compact groups, *Proc. London Math. Soc.*, **45**, 506–518, 1982.

[SS] T. Slaman and J. Steel, Definable functions on degrees, *Cabal Seminar 81–85*, Lecture Notes in Math., **1333**, Springer-Verlag, 37–55, 1988.

[SWW] D. Sullivan, B. Weiss, and J.B. Wright, Generic dynamics and monotone complete C^*-algebras, *Trans. Amer. Math. Soc.*, **295**, 795–809, 1986.

[T01] S. Thomas, On the complexity of the classification problem for torsion-free abelian groups of finite rank, *Bull. Symb. Logic*, **7(3)**, 329–344, 2001.

[T02] S. Thomas, The classification problem for p-local torsion-free abelian groups of finite rank, preprint, 2002.

[T02a] S. Thomas, Some applications of superrigidity to Borel equivalence relations, *Set Theory*, S. Thomas (Ed.), DIMACS Series, **58**, Amer. Math. Soc., 129–134, 2002.

[V] V.S. Varadarajan, Groups of automorphisms of Borel spaces, *Trans. Amer. Math. Soc.*, **109**, 191–220, 1963.

[W84] B. Weiss, Measurable dynamics, *Contemp. Math.*, **26**, 395–421, 1984.

[W00] B. Weiss, A survey of generic dynamics, *Descriptive Set Theory and Dynamical Systems*, Ed. by M. Foreman et al., London Math. Soc. Lecture Note Series, **277**, Cambridge Univ. Press, 273–291, 2000.

[Zi78] R. Zimmer, Amenable ergodic group actions and an application to Poisson boundaries of random walks, *J. Funct. Anal.*, **27**, 350–372, 1978.

[Zi81] R. Zimmer, Orbit equivalence and rigidity of ergodic actions of Lie groups, *Erg. Th. and Dynam. Syst.*, **1**, 237–253, 1981.

[Zi84] R. Zimmer, *Ergodic Theory and Semisimple Groups*, Birkhäuser, 1984.

Editorial Information

To be published in the *Memoirs*, a paper must be correct, new, nontrivial, and significant. Further, it must be well written and of interest to a substantial number of mathematicians. Piecemeal results, such as an inconclusive step toward an unproved major theorem or a minor variation on a known result, are in general not acceptable for publication. Papers appearing in *Memoirs* are generally at least 80 and not more than 200 published pages in length. Papers less than 80 or more than 200 published pages require the approval of the Managing Editor of the Transactions/Memoirs Editorial Board.

As of May 31, 2005, the backlog for this journal was approximately 11 volumes. This estimate is the result of dividing the number of manuscripts for this journal in the Providence office that have not yet gone to the printer on the above date by the average number of monographs per volume over the previous twelve months, reduced by the number of volumes published in four months (the time necessary for preparing a volume for the printer). (There are 6 volumes per year, each containing at least 4 numbers.)

A Consent to Publish and Copyright Agreement is required before a paper will be published in the *Memoirs*. After a paper is accepted for publication, the Providence office will send a Consent to Publish and Copyright Agreement to all authors of the paper. By submitting a paper to the *Memoirs*, authors certify that the results have not been submitted to nor are they under consideration for publication by another journal, conference proceedings, or similar publication.

Information for Authors

Memoirs are printed from camera copy fully prepared by the author. This means that the finished book will look exactly like the copy submitted.

The paper must contain a *descriptive title* and an *abstract* that summarizes the article in language suitable for workers in the general field (algebra, analysis, etc.). The *descriptive title* should be short, but informative; useless or vague phrases such as "some remarks about" or "concerning" should be avoided. The *abstract* should be at least one complete sentence, and at most 300 words. Included with the footnotes to the paper should be the 2000 *Mathematics Subject Classification* representing the primary and secondary subjects of the article. The classifications are accessible from `www.ams.org/msc/`. The list of classifications is also available in print starting with the 1999 annual index of *Mathematical Reviews*. The Mathematics Subject Classification footnote may be followed by a list of *key words and phrases* describing the subject matter of the article and taken from it. Journal abbreviations used in bibliographies are listed in the latest *Mathematical Reviews* annual index. The series abbreviations are also accessible from `www.ams.org/publications/`. To help in preparing and verifying references, the AMS offers MR Lookup, a Reference Tool for Linking, at `www.ams.org/mrlookup/`. When the manuscript is submitted, authors should supply the editor with electronic addresses if available. These will be printed after the postal address at the end of the article.

Electronically prepared manuscripts. The AMS encourages electronically prepared manuscripts, with a strong preference for $\mathcal{A}_{\mathcal{M}}\mathcal{S}$-LaTeX. To this end, the Society has prepared $\mathcal{A}_{\mathcal{M}}\mathcal{S}$-LaTeX author packages for each AMS publication. Author packages include instructions for preparing electronic manuscripts, the *AMS Author Handbook*, samples, and a style file that generates the particular design specifications of that publication series. Though $\mathcal{A}_{\mathcal{M}}\mathcal{S}$-LaTeX is the highly preferred format of TeX, author packages are also available in $\mathcal{A}_{\mathcal{M}}\mathcal{S}$-TeX.

Authors may retrieve an author package from e-MATH starting from `www.ams.org/tex/` or via FTP to `ftp.ams.org` (login as `anonymous`, enter username as password, and type `cd pub/author-info`). The *AMS Author Handbook* and the *Instruction Manual* are available in PDF format following the author packages link from `www.ams.org/tex/`. The author package can be obtained free of charge by sending email to pub@ams.org (Internet) or from the Publication Division, American Mathematical Society, 201 Charles St., Providence, RI 02904, USA. When requesting an author package, please specify $\mathcal{A}\mathcal{M}\mathcal{S}$-LaTeX or $\mathcal{A}\mathcal{M}\mathcal{S}$-TeX, Macintosh or IBM (3.5) format, and the publication in which your paper will appear. Please be sure to include your complete mailing address.

Sending electronic files. After acceptance, the source file(s) should be sent to the Providence office (this includes any TeX source file, any graphics files, and the DVI or PostScript file).

Before sending the source file, be sure you have proofread your paper carefully. The files you send must be the EXACT files used to generate the proof copy that was accepted for publication. For all publications, authors are required to send a printed copy of their paper, which exactly matches the copy approved for publication, along with any graphics that will appear in the paper.

TeX files may be submitted by email, FTP, or on diskette. The DVI file(s) and PostScript files should be submitted only by FTP or on diskette unless they are encoded properly to submit through email. (DVI files are binary and PostScript files tend to be very large.)

Electronically prepared manuscripts can be sent via email to pub-submit@ams.org (Internet). The subject line of the message should include the publication code to identify it as a Memoir. TeX source files, DVI files, and PostScript files can be transferred over the Internet by FTP to the Internet node e-math.ams.org (130.44.1.100).

Electronic graphics. Comprehensive instructions on preparing graphics are available at www.ams.org/jourhtml/graphics.html. A few of the major requirements are given here.

Submit files for graphics as EPS (Encapsulated PostScript) files. This includes graphics originated via a graphics application as well as scanned photographs or other computer-generated images. If this is not possible, TIFF files are acceptable as long as they can be opened in Adobe Photoshop or Illustrator. No matter what method was used to produce the graphic, it is necessary to provide a paper copy to the AMS.

Authors using graphics packages for the creation of electronic art should also avoid the use of any lines thinner than 0.5 points in width. Many graphics packages allow the user to specify a "hairline" for a very thin line. Hairlines often look acceptable when proofed on a typical laser printer. However, when produced on a high-resolution laser imagesetter, hairlines become nearly invisible and will be lost entirely in the final printing process.

Screens should be set to values between 15% and 85%. Screens which fall outside of this range are too light or too dark to print correctly. Variations of screens within a graphic should be no less than 10%.

Inquiries. Any inquiries concerning a paper that has been accepted for publication should be sent directly to the Electronic Prepress Department, American Mathematical Society, 201 Charles St., Providence, RI 02904, USA.

Titles in This Series

836 **H. G. Dales and A. T.-M. Lau,** The second duals of Beurling algebras, 2005

835 **Kiyoshi Igusa,** Higher complex torsion and the framing principle, 2005

834 **Ken'ichi Ohshika,** Kleinian groups which are limits of geometrically finite groups, 2005

833 **Greg Hjorth and Alexander S. Kechris,** Rigidity theorems for actions of product groups and countable Borel equivalence relations, 2005

832 **Lee Klingler and Lawrence S. Levy,** Representation type of commutative Noetherian rings III: Global wildness and tameness, 2005

831 **K. R. Goodearl and F. Wehrung,** The complete dimension theory of partially ordered systems with equivalence and orthogonality, 2005

830 **Jason Fulman, Peter M. Neumann, and Cheryl E. Praeger,** A generating function approach to the enumeration of matrices in classical groups over finite fields, 2005

829 **S. G. Bobkov and B. Zegarlinski,** Entropy bounds and isoperimetry, 2005

828 **Joel Berman and Paweł M. Idziak,** Generative complexity in algebra, 2005

827 **Trevor A. Welsh,** Fermionic expressions for minimal model Virasoro characters, 2005

826 **Guy Métivier and Kevin Zumbrun,** Large viscous boundary layers for noncharacteristic nonlinear hyperbolic problems, 2005

825 **Yaozhong Hu,** Integral transformations and anticipative calculus for fractional Brownian motions, 2005

824 **Luen-Chau Li and Serge Parmentier,** On dynamical Poisson groupoids I, 2005

823 **Claus Mokler,** An analogue of a reductive algebraic monoid whose unit group is a Kac-Moody group, 2005

822 **Stefano Pigola, Marco Rigoli, and Alberto G. Setti,** Maximum principles on Riemannian manifolds and applications, 2005

821 **Nicole Bopp and Hubert Rubenthaler,** Local zeta functions attached to the minimal spherical series for a class of symmetric spaces, 2005

820 **Vadim A. Kaimanovich and Mikhail Lyubich,** Conformal and harmonic measures on laminations associated with rational maps, 2005

819 **F. Andreatta and E. Z. Goren,** Hilbert modular forms: Mod p and p-adic aspects, 2005

818 **Tom De Medts,** An algebraic structure for Moufang quadrangles, 2005

817 **Javier Fernández de Bobadilla,** Moduli spaces of polynomials in two variables, 2005

816 **Francis Clarke,** Necessary conditions in dynamic optimization, 2005

815 **Martin Bendersky and Donald M. Davis,** V_1-periodic homotopy groups of $SO(n)$, 2004

814 **Johannes Huebschmann,** Kähler spaces, nilpotent orbits, and singular reduction, 2004

813 **Jeff Groah and Blake Temple,** Shock-wave solutions of the Einstein equations with perfect fluid sources: Existence and consistency by a locally inertial Glimm scheme, 2004

812 **Richard D. Canary and Darryl McCullough,** Homotopy equivalences of 3-manifolds and deformation theory of Kleinian groups, 2004

811 **Ottmar Loos and Erhard Neher,** Locally finite root systems, 2004

810 **W. N. Everitt and L. Markus,** Infinite dimensional complex symplectic spaces, 2004

809 **J. T. Cox, D. A. Dawson, and A. Greven,** Mutually catalytic super branching random walks: Large finite systems and renormalization analysis, 2004

808 **Hagen Meltzer,** Exceptional vector bundles, tilting sheaves and tilting complexes for weighted projective lines, 2004

807 **Carlos A. Cabrelli, Christopher Heil, and Ursula M. Molter,** Self-similarity and multiwavelets in higher dimensions, 2004

806 **Spiros A. Argyros and Andreas Tolias,** Methods in the theory of hereditarily indecomposable Banach spaces, 2004